INTRODUCTION TO

HIGH ENERGY PHYSICS

Particle Physics for the Beginner

Problems and Solutions

INTRODUCTION TO
HIGH ENERGY PHYSICS
Particle Physics for the Beginner

Problems and Solutions

Lee G Pondrom
University of Wisconsin–Madison, USA

W& World Scientific

NEW JERSEY · LONDON · SINGAPORE · BEIJING · SHANGHAI · HONG KONG · TAIPEI · CHENNAI · TOKYO

Published by

World Scientific Publishing Co. Pte. Ltd.

5 Toh Tuck Link, Singapore 596224

USA office: 27 Warren Street, Suite 401-402, Hackensack, NJ 07601

UK office: 57 Shelton Street, Covent Garden, London WC2H 9HE

British Library Cataloguing-in-Publication Data
A catalogue record for this book is available from the British Library.

INTRODUCTION TO HIGH ENERGY PHYSICS: PARTICLE PHYSICS FOR THE BEGINNER
Problems and Solutions

ISBN 978-981-127-012-3 (hardcover)
ISBN 978-981-127-032-1 (paperback)
ISBN 978-981-127-013-0 (ebook for institutions)
ISBN 978-981-127-014-7 (ebook for individuals)

For any available supplementary material, please visit
https://www.worldscientific.com/worldscibooks/10.1142/13243#t=suppl

Typeset by Stallion Press
Email: enquiries@stallionpress.com

Preface to Solutions

At first the Editors at World Scientific suggested that it would be useful to post the solutions to the problems online. The author agreed to this proposal, and six months later we had the solutions. It was then decided to publish the Solutions Manual as a separate book. This is reasonable, as the Manual is a considerable resource, although it is not a stand-alone object. For clarity, the problem statements from the text are repeated here, but several references to equations in the text remain, so this book has to be viewed as a companion to *"Introduction to High Energy Physics"*. The two must be used side by side.

The good news is that there are no problems that are impossible to solve. Another advantage of the exercise is that the text, and in particular, the problem sets, got a thoroughgoing check for errors. Indeed, some were found, and are corrected in the following reprints. In the original text they are:

Chapter 1, Problem 1.8 the proton spin wave function should have a minus sign in the middle:

$$p \uparrow = \sqrt{2/3} d \downarrow u \uparrow u \uparrow - (d \uparrow (u \uparrow u \downarrow + u \downarrow u \uparrow)) / \sqrt{6}$$

Problem 1.10, spelling error isosceles has no final e.

Problem 1.11, Equation (1.38), not (1.26).

Chapter 7, Problem 7.1, polarization in Møller scattering, in Equation (7.147), the algebraic signs multiplying $\lambda_1 \lambda_2$ should all be $+$.

This turns Equation (7.148) upside down! It is the $m_s = 0$ cross-section that is the larger.

Chapter 8, Problem 8.1, minus sign is missing in Equation (8.111). Equation (3.207) is correct.

Problem 8.5, in checking Equation (8.25), a sign error was found in the middle term. It should be $+g\epsilon_{ijk}W_{j\nu}W_k^{\nu\sigma}$.

Chapter 9, Problem 9.12 Equation (9.83) instead of (9.82).

Any other errors found by users of the text should be sent to World Scientific and to the author pondrom@hep.wisc.edu

Contents

Chapter 1

Problem 1.1. Using the Clebsch–Gordan coefficients for $I = 1/2$ plus $I = 1$, obtain the amplitudes displayed in Equations (1.2)–(1.4).

Solution 1.1. Write down the kets for $|\pi p\rangle$ using the Clebsch–Gordan coefficients of Table 1.2. Recall that the C–G coefficients are a real unitary transformation, so the same matrix element goes $|I_1, I_2\rangle \rightarrow |I, I_z\rangle$ as $|I, I_z\rangle \rightarrow |I_1, I_2\rangle$.

$$\left|p\pi^+\right\rangle = |I_1 = 1/2, I_2 = 1; +1/2, +1\rangle = |I = 3/2, I_z = +3/2\rangle\,;$$

$$\left|p\pi^-\right\rangle = |I_1 = 1/2, I_2 = 1; +1/2, -1\rangle$$

$$= \frac{(|I = 3/2, I_z = -1/2\rangle - \sqrt{2}\,|I = 1/2, I_z = -1/2\rangle)}{\sqrt{3}}\,;$$

$$\left|n\pi^0\right\rangle = |I_1 = 1/2, I_2 = 1; -1/2, 0\rangle$$

$$= \frac{(\sqrt{2}\,|I = 3/2, I_z = -1/2\rangle + |I = 1/2, I_z = -1/2\rangle)}{\sqrt{3}}.$$

Then calculate the amplitudes, keeping in mind that the $I = 3/2$ and $I = 1/2$ states are orthogonal:

$$\langle \pi^+ p|\, H_{3/2} + H_{1/2}\, |\pi^+ p\rangle = A_{3/2}\,;$$

$$\langle \pi^- p|\, H_{3/2} + H_{1/2}\, |\pi^- p\rangle = \frac{A_{3/2} + 2A_{1/2}}{3}\,;$$

$$\langle \pi^- p|\, H_{3/2} + H_{1/2}\, |\pi^0 n\rangle = \frac{\sqrt{2}(A_{3/2} - A_{1/2})}{3}.$$

The magnitude squares of these amplitudes give Equations (1.2), (1.3), and (1.4).

Problem 1.2. The deuteron has I-spin zero, and this leads to several easy tests of charge independence. Show that I-spin conservation predicts:

a. $\sigma(\pi^+ + d \to p + p)/\sigma(\pi^- + d \to n + n) = 1$.
b. $\sigma(p + d \to \mathrm{He}^3 + \pi^0)/\sigma(p + d \to \mathrm{H}^3 + \pi^+) = 1/2$ (He^3 and H^3 are an I-spin doublet).
c. $\sigma(d + d \to \mathrm{He}^4 + \pi^0) = 0$ (He^4 has I-spin zero).

Solution 1.2. The hypothesis of charge independence means that the reactions depend on I, but not I_z, so $\sigma(\pi^+ d \to p + p) = \sigma(\pi^- d \to n + n)$. The initial state $p + d$ must have $I = 1/2$, because the deuteron has $I = 0$. He^3 and H^3 are an I-spin doublet with $I_z = +1/2$ and $I_z = -1/2$ respectively.

$$\left| He^3\pi^0 \right\rangle = |I_1 = 1/2, I_2 = 1; +1/2, 0\rangle$$

$$= \sqrt{\frac{2}{3}} |I = 3/2, I_z = +1/2\rangle - \sqrt{\frac{1}{3}} |I = 1/2, I_z = +1/2\rangle ;$$

and

$$\left| H^3\pi^+ \right\rangle = |I_1 = 1/2, I_2 = 1; -1/2, +1\rangle$$

$$= \sqrt{\frac{1}{3}} |I = 3/2, I_z = +1/2\rangle + \sqrt{\frac{2}{3}} |I = 1/2, I_z = +1/2\rangle ;$$

leading to the prediction that the $H^3\pi^+$ cross-section is twice as large as $He^3\pi^0$. For reaction (c), the π^0 has I-spin one, so the final state is forbidden.

Problem 1.3. Strange particle reactions. The K's are doublets, and the Σ's are triplets, with the Λ a singlet. Obtain the following relations:

a. $\langle \pi^+ p | H_{3/2} + H_{1/2} | \Sigma^+ K^+ \rangle = A_{3/2}$.
b. $\langle \pi^- p | H_{3/2} + H_{1/2} | \Sigma^- K^+ \rangle = 1/\sqrt{3}(A_{3/2} + 2A_{1/2})$
c. $\langle \pi^- p | H_{3/2} + H_{1/2} | \Sigma^0 K^0 \rangle = \sqrt{2}(A_{3/2} - A_{1/2})/\sqrt{3}$.

Identical in form to π p scattering, except the amplitudes are not the same.

d. $\langle \bar{K}^0 p | H_1 + H_0 | \Lambda \pi^+ \rangle = A_1$.

e. $\langle \bar{K}^0 n | H_1 + H_0 | \Lambda \pi^0 \rangle = A_1 / \sqrt{2}$.

Solution 1.3. For the strange particles the mesons and baryons switch roles: K's have $I = 1/2$ and Σ's have $I = 1$, with $\Lambda I = 0$ and $\Xi I = 1/2$. The amplitudes are:

$$\left| K^+ \Sigma^+ \right\rangle = |I_1 = 1/2, I_2 = 1; +1/2, +1\rangle = |I = 3/2, I_z = +3/2\rangle ;$$

$$\left| K^+ \Sigma^- \right\rangle = |I_1 = 1/2, I_2 = 1; +1/2, -1\rangle$$

$$= \sqrt{\frac{1}{3}} |I = 3/2; I_z = -1/2\rangle - \sqrt{\frac{2}{3}} |I = 1/2, I_z = -1/2\rangle$$

$$\left| K^0 \Sigma^0 \right\rangle = |I_1 = 1/2, I_2 = 1; -1/2, 0\rangle$$

$$= \sqrt{\frac{2}{3}} |I = 3/2, I_z = -1/2\rangle + \sqrt{\frac{1}{3}} |I = 1/2, I_z = -1/2\rangle .$$

These kets, plus those in problem 1.1 give the amplitudes (a), (b), and (c). For (d) and (e), $\Lambda \pi$ has $I = 1$, so the $I_1 = 1/2$, $I_2 = 1/2$ C–G coefficients come into play. The initial states are: $\left| \bar{K}^0 p \right\rangle = |I_1 = 1/2, I_2 = 1/2; +1/2, +1/2\rangle$ and $\left| \bar{K}^0 n \right\rangle = |I_1 = 1/2, I_2 = 1/2; +1/2, -1/2\rangle$.

Problem 1.4. *G* parity. This is a combination of charge conjugation and a rotation of 180° about the y axis in I-spin space: $G = C \times e^{i\pi I_y}$. An even number of pions has even G parity, and an odd number odd. For $I = 1$, the I_y matrix is

$$\begin{pmatrix} 0 & -i/\sqrt{2} & 0 \\ i/\sqrt{2} & 0 & -i/\sqrt{2} \\ 0 & i/\sqrt{2} & 0 \end{pmatrix} .$$

Show that $I_y^3 = I_y$. This does not mean that $I_y^2 = 1$, because I_y is a singular matrix, as are all matrices for integer I-spin. This polynomial can be factored: $I_y(I_y^2 - 1) = 0$, with roots that are the eigenvalues ± 1 and 0. This is a property of all of the I-spin representations. For $I = 1/2$, $I_y^2 - 1/4 = 0$, and for $I = 3/2$, $(I_y^2 - 9/4)(I_y^2 - 1/4) = 0$.

All three operators I_x, I_y, and I_z have the same eigenvalues, and obey the same polynomial equation in a given representation. Now show that for $I = 1$:

$$e^{i\theta I_y} = 1 + (\cos\theta - 1)I_y^2 + i\sin\theta I_y. \qquad (1.1)$$

Then

$$e^{i\pi I_y} = 1 - 2I_y^2 \qquad (1.2)$$

which is the matrix:

$$\begin{pmatrix} 0 & 0 & 1 \\ 0 & -1 & 0 \\ 1 & 0 & 0 \end{pmatrix} \begin{pmatrix} \pi^+ \\ \pi^0 \\ \pi^- \end{pmatrix} = \begin{pmatrix} \pi^- \\ -\pi^0 \\ \pi^+ \end{pmatrix} \qquad (1.3)$$

Now because $\pi^0 \to \gamma + \gamma$, $C|\pi^0\rangle = +|\pi^0\rangle$, so for consistency $C|\pi^+\rangle = -|\pi^-\rangle$. Then $G|\pi\rangle = -|\pi\rangle$ for all three charge states.

Solution   1.4. G Parity problem. Matrix multiplication gives $I_y^3 = I_y$:

$$I_y = \begin{pmatrix} 0 & -i/\sqrt{2} & 0 \\ i/\sqrt{2} & 0 & -i/\sqrt{2} \\ 0 & i/\sqrt{2} & 0 \end{pmatrix}; \quad I_y^2 = \begin{pmatrix} 1/2 & 0 & -1/2 \\ 0 & 1 & 0 \\ -1/2 & 0 & 1/2 \end{pmatrix}; \quad I_y^3 = I_y.$$

For the second part expand $e^{i\theta I_y}$ and collect terms:

$$\begin{aligned}
e^{i\theta I_y} &= 1 + i\theta I_y + (i\theta I_y)^2/(2!) + (i\theta I_y)^3/(3!) + (i\theta I_y)^4/(4!) \\
&\quad + (i\theta I_y)^5/(5!) + \cdots \\
&= 1 + (-\theta^2/2! + \theta^4/4! - \theta^6/6! + \cdots)I_y^2 \\
&\quad + i(\theta - \theta^3/3! + \theta^5/5! + \cdots)I_y;
\end{aligned}$$

gives Equation (1.1):

$$e^{i\theta I_y} = 1 + (\cos\theta - 1)I_y^2 + i\sin\theta I_y; \quad e^{i\pi I_y} = 1 - 2I_y^2 = \begin{pmatrix} 0 & 0 & 1 \\ 0 & -1 & 0 \\ 1 & 0 & 0 \end{pmatrix}.$$

Then defining the G parity as $G = Ce^{i\pi I_y}$ and $C|\pi^+\rangle = -|\pi^-\rangle$, $C|\pi^0\rangle = +|\pi^0\rangle$ gives $G|n\,\pi\rangle = (-1)^n |n\pi\rangle$.

Problem 1.5. Evaluate the scattering amplitude $f(\theta)$ — Equation (1.5) — for S and P waves only, i.e. $l = 0$ and $l = 1$. Show that the angular distribution has the form:

$$d\sigma/d\Omega = |f(\theta)|^2 = A + B \times \cos\theta + C \times \cos^2\theta. \qquad (1.4)$$

Evaluate the coefficients A, B, and C in terms of δ_0 and δ_1. Now assume $\delta_0 = 0$, and $\delta_1 = \pi/2$, and calculate the total cross-section for relative momentum $\hbar k = 265\,\text{MeV}/c$. Compare your result with the peak value at 1235 MeV in Figure 1.1. The formula $1\,\text{GeV}^{-2} = 0.389\,\text{mb}$ is useful here. There is a small error in that $(2l+1)$ should be replaced by $(2j+1)$, or multiply by four instead of three.

Solution 1.5. For S and P waves,

$$|f(\theta)|^2 = \frac{1}{k^2}(\sin^2\delta_0 + 6\sin\delta_0\sin\delta_1\cos(\delta_1 - \delta_0)\cos\theta + 9\sin^2\delta_1\cos^2\theta).$$

For $\delta_0 = 0$ and $\delta_1 = \pi/2$ and $k = 265\,\text{MeV}/c$:

$$\sigma_{tot} = \frac{12\pi}{k^2} = 200\,\text{mb}.$$

This is the peak value of the cross section near $0.3\,\text{GeV}/c$ in the top plot of Figure 1.1.

Problem 1.6. Practice with the eight Gell-Mann matrices and I-spin, V-spin, U-spin. Work out some commutation rules:

a. $[I_z, V_\pm] = \pm V_\pm/2$;
b. $[U_+, U_-] = 3Y/2 - I_z \equiv 2U_z$;
c. $[V_+, V_-] = 3Y/2 + I_z \equiv 2V_z$;
d. $[U_+, V_-] = I_-$.

Solution 1.6. Use $[\lambda_i/2, \lambda_j/2] = if_{ijk}\lambda_k/2$, and the totally anti-symmetric structure constants f_{ijk} in Table 1.3, together with the definitions in Equation 1.22:

$$I_\pm = (\lambda_1 \pm i\lambda_2)/2; \quad U_\pm = (\lambda_6 \pm i\lambda_7)/2; \quad V_\pm = (\lambda_4 \pm i\lambda_5)/2;$$

$$I_z = \lambda_3/2; \quad Y = \lambda_8/\sqrt{3}.$$

$$I_z = V_z - U_z; \quad V_z = (3Y + 2I_z)/4 = (\sqrt{3}\lambda_8 + \lambda_3)/4;$$

$$U_z = (\sqrt{3}\lambda_8 - \lambda_3)/4.$$

We will work out some examples:

$$[I_z, V_\pm] = [\lambda_3/2, \lambda_4/2] \pm i[\lambda_3/2, \lambda_5/2] = (\lambda_4 \pm i\lambda_5)/4 = V_\pm/2;$$

and

$$[U_+, U_-] = [\lambda_6 + i\lambda_7, \lambda_6 - i\lambda_7]/4 = -i[\lambda_6, \lambda_7]/2$$

$$= -2i[\lambda_6/2, \lambda_7/2] = -\lambda_3/2 + \sqrt{3}\lambda_8/2.$$

Problem 1.7. Check the matrix elements for $\langle \pi^0| V_- |K^+\rangle = 1/\sqrt{2}$ and $\langle \pi^0| U_- |K^0\rangle = -1/\sqrt{2}$.

Solution 1.7. Octet matrix elements for $\langle \pi^0| V_- |K^+\rangle$, and $\langle \pi^0| U_- |K^0\rangle$. Refer to Figure 1.7 for the $U_\pm$ and $V_\pm$ octet operators. K^+ and K^0 are at the top, and π^0 together with η^0 are in the center. The convention adopted for the quarks is that all three matrix elements for $\langle u| V_- |s\rangle$ etc. equal $+1$, and for the antiquarks only $\langle \bar{s}| U_+ |\bar{d}\rangle$ equals -1. So we may operate on the quark content of the K mesons:

$$V_- |K^+\rangle = V_- |u\bar{s}\rangle = |s\bar{s}\rangle + |u\bar{u}\rangle ;$$

$$\langle \pi^0| V_- |K^+\rangle = (\langle u\bar{u}| + \langle d\bar{d}|)(|s\bar{s}\rangle + |u\bar{u}\rangle)/\sqrt{2} = 1/\sqrt{2}.$$

The K^0 analysis is the same, except a minus sign appears on the antiquarks due to the above convention.

Problem 1.8. Use color symmetry to show that the baryons around the outside of the octet must have the like quarks in a triplet spin state. Define the baryon magnetic moment:

$\langle \vec{\mu_B}\rangle = \langle S = 1/2, Sz = +1/2| \sum_q \vec{\mu_q}|S = 1/2, Sz = +1/2\rangle$, where the sum is over the three quark magnetic moments. The magnetic moment is written as

$$\vec{\mu_q} = g_q \times (q/2m_q) \times \vec{S},$$

where the Lande g factor is defined by this equation, and $q = e_q \times e$.
    Obtain the proton spin wave function:

$$p\uparrow= \sqrt{2/3}\, d \downarrow u \uparrow u \uparrow -(d \uparrow (u \uparrow u \downarrow +u \downarrow u \uparrow))/\sqrt{6},$$

in an obvious notation. Calculate the z component of the magnetic moment, and show that $\mu_p = 4/3\mu_u - \mu_d/3$. This formula is true for all octet baryons with two like flavor quarks: $\mu_B = 4/3\mu_2 - 1/3\mu_1$. Use this, plus $\mu_u = -2\mu_d$ (same mass and g factor, charges $e_q = +2/3$ and $-1/3$) to show that $\mu_p/\mu_n = -3/2$. Look up the numbers and compare with experiment. From the values of μ_p and μ_n calculate the quark moments μ_u and μ_d. The baryons in the middle of the octet Σ^0 and Λ have the same quark content (uds). Assume orthogonality, and (ud) in a triplet spin state for Σ^0 matching (uu) and (dd) for Σ^+ and Σ^-. Then the spin and magnetic moment of the Λ comes only from the strange quark. Use the measured value of μ_Λ to obtain μ_s, and then calculate μ_{Σ^+}, μ_{Σ^-}, μ_{Ξ^0}, and μ_{Ξ^-}. Compare results with experiment.

Solution 1.8. Look at the proton. The two u quarks must be anti-symmetric in the exchange of all quantum numbers to obey the Pauli exclusion principle. There are three colors, and $3 \times 3 = 6 + 3^*$. The 3^* couples to the 3 colors of the d quark to form a singlet color-less baryon state. $(3 \times 3^* = 8 + 1)$. The 6 is symmetric under the exchange of the two quarks, so the 3^* must be antisymmetric to be orthogonal. Color takes care of the Pauli principle, so the spin state has to be even, that is $S = 1$. With this information, and the handy C–G Table 1.2, the wave function for a proton with spin up is:

$$p\uparrow= \sqrt{\frac{2}{3}}d\downarrow u\uparrow u\uparrow -(u\uparrow u\downarrow +u\downarrow u\uparrow)\frac{d\uparrow}{\sqrt{6}}.$$

The minus sign agrees with the phase choice in Table 1.2. Then the proton magnetic moment in the quark model is:

$$\left\langle p\uparrow\left|\sum_q \vec{\mu}_q\right|p\uparrow\right\rangle = \frac{2}{3}(2\mu_u - \mu_d) + \frac{2}{6}\mu_d = \frac{4\mu_u - \mu_d}{3}.$$

The $m_s = 0$ diquark adds a factor of two, but no net magnetic moment. This formula is correct for all the baryons around the outside of the octet. The Λ in the center has the magnetic moment of the strange quark, because the up and down quarks are in a singlet state. Assuming $m_u = m_d$, the ratio $\mu_u/\mu_d = -2$, so in the quark model:

$$\frac{\mu_p}{\mu_n} = \frac{4\mu_u - \mu_d}{4\mu_d - \mu_u} = -\frac{9}{6} = -1.5; \quad \text{measured numbers} \quad -\frac{2.79}{1.91} = -1.46.$$

The measured values are in units of the nuclear magneton $\mu_n = e\hbar/(2M_pc)$. In these units we can solve for the up quark magnetic moment from the proton moment:

$$\mu_p = (4/3 + 1/6)\mu_u; \text{ or } \mu_u = \frac{18}{27}\mu_p = 1.86 \text{ nuclear magnetons.}$$

Assuming that the up quark g factor is two, and the charge is $+2/3$ giving a mass ratio $m_u/M_p = 0.36$, close to expectations for a constituent quark mass. The Particle Data Group lists the baryon magnetic moments, that are compared to our quark model calculations in the following Table:

Baryon Magnetic Moments

Name	μ in nuclear magnetons	quark model
Λ	-0.613 ± 0.004	input
Σ^+	2.458 ± 0.01	$+2.68$
Σ^-	-1.160 ± 0.025	-1.04
Ξ^0	-1.250 ± 0.014	-1.43
Ξ^-	-0.6507 ± 0.0025	-0.50
Ω^-	-2.02 ± 0.05	-1.83

So with no special effort to optimize the numbers, the agreement with the quark model assuming $\mu_d = -\mu_u/2$ and $\mu_s = \mu_\Lambda$ is good to about 20% — not too bad. A prediction of the model is that all of the neutral baryons have negative magnetic moments, consistent with experiment. The assumption $\mu_d = -\mu_u/2$ can be relaxed by solving the proton and neutron moment equations for μ_u and μ_d separately. This gives slightly different numbers. For the strange quark one could solve μ_{Σ^+} or μ_{Σ^-} for μ_s, or try $\mu_s = \mu_\Omega/3$, instead of $\mu_s = \mu_\Lambda$; or attempt an overall fit of three quark moments to eight measured baryon moments. Nothing works perfectly, but all of the fits are reasonable.

The Σ^0 is left out. It has a short lifetime for electromagnetic decay $\Sigma^0 \to \Lambda + \gamma$, so the quantity that can be measured is the transition magnetic moment $\langle \Sigma^0 \uparrow | (\vec{\mu}_u + \vec{\mu}_d + \vec{\mu}_s) \cdot \hat{z} | \Lambda \uparrow \rangle$. The measured quantity is the square of this matrix element, that is proportional to the cross-section for Σ^0 production in a Λ beam via one photon exchange — a process called the Primakoff effect after Henry Primakoff who first proposed it. P.C. Petersen *et al.*, *Phys Rev Lett* **57**, 949, (1986) quoted a number 1.59±0.05±0.07 nuclear magnitons. This matrix element can be calculated in our quark model:

$$(\mu_{uz} + \mu_{dz} + \mu_{sz}) |\Lambda \uparrow\rangle$$
$$= (\mu_u(u\uparrow d\downarrow + u\downarrow d\uparrow)s\uparrow - \mu_d(u\uparrow d\downarrow + d\uparrow u\downarrow)s\uparrow$$
$$+ \mu_s(u\uparrow d\downarrow - u\downarrow d\uparrow)s\uparrow)/\sqrt{2}.$$

The quark content of the Σ^0 is:

$$|\Sigma^0 \uparrow\rangle = \sqrt{\frac{2}{3}}s\downarrow u\uparrow d\uparrow - (u\uparrow d\downarrow + u\downarrow d\uparrow)s\uparrow /\sqrt{6}.$$

Hence

$$\langle\Sigma^0 \uparrow| (\mu_{uz} + \mu_{dz} + \mu_{sz}) |\Lambda \uparrow\rangle = \frac{1}{\sqrt{3}}(\mu_d - \mu_u) = -1.6.$$

This is consistent with the magnitude quoted by Petersen *et al.* The sign has not been measured.

Problem 1.9. $\Delta Q = \Delta S$ rule. This rule applies to the weak decays of strange particles. We have seen how $\Delta Q = \Delta S$ forbids the beta decay $\Sigma^+ \to n + e^+ + \nu_e$ ($\Delta Q = -1, \Delta S = +1$). Apply the $\Delta Q = \Delta S$ rule to the following:

a. $\Xi^0 \to \Sigma^+ + e^- + \bar{\nu}_e$;
b. $\Xi^0 \to \Sigma^- + e^+ + \nu_e$;
c. $K^0 \to \pi^+ + e^- + \bar{\nu}_e$;
d. $K^0 \to \pi^- + e^+ + \nu_e$.

Solution 1.9. Draw the baryon and meson octets to understand the $\Delta Q = \Delta S$ rule. If the states can be connected by a single SU_3 operator, then $\Delta Q = \Delta S$; otherwise no. Hence $a = $ yes, $b = $ no, $c = $ no, and $d = $ yes. $\bar{K}^0 \to \pi^+$ is OK. The decay rates for $K^0 \to \pi^-$ and $\bar{K}^0 \to \pi^+$ are slightly different, that is a CP violation discussed in Chapter 9.

Problem 1.10. $|\Delta I| = 1/2$ rule. This rule also applies to weak decays of strange particles. For some tests it is useful to use a 'spurion', a particle that has only I-spin $1/2$ and nothing else (no energy, spin, etc.). Add in a spurion, and conserve I-spin, and viola! You have the $|\Delta I| = 1/2$ rule. Start with some easy ones. w is the decay rate. Show that the following are consequences of the $|\Delta I| = 1/2$ rule, and compare with experiment:

a. $w(\Lambda \to p + \pi^-)/w(\Lambda \to n + \pi^0) = 2$.

b. $w(K^0 \to \pi^+\pi^-)/w(K^0 \to \pi^0\pi^0) = 2$.

c. Add a spurion to Σ^+ and Σ^- to obtain the amplitudes:

1. $A(\mathrm{sp} + \Sigma^+ \to p + \pi^0) = \sqrt{2}(A_{3/2} + A_{1/2})/3$;

2. $A(\mathrm{sp} + \Sigma^+ \to n + \pi^+) = (A_{3/2} - 2A_{1/2})/3$;

3. $A(\mathrm{sp} + \Sigma^- \to n + \pi^-) = A_{3/2}$.

These are complex numbers that satisfy a 'triangular relation' $\sqrt{2}A_0 + A_+ - A_- = 0$, where the subscript refers to the charge of the pion. Look up the numbers. The magnitudes of the three amplitudes are nearly the same: $|A_0| = |A_+| = |A_-|$, so it is an isosceles right triangle, with $\sqrt{2}A_0$ the hypotenuse.

Solution 1.10. $|\Delta I| = 1/2$ rule. The Λ decay is easy, because $I = 0$, so the $p\pi^-$ and $n\pi^0$ must both have $I = 1/2$.

$$\left|p\pi^-\right\rangle = \left|I_1 = 1/2, I_2 = 1; +1/2, -1\right\rangle$$

$$= \sqrt{\frac{1}{3}}\left|3/2; -1/2\right\rangle - \sqrt{\frac{2}{3}}\left|1/2, -1/2\right\rangle ;$$

$$\left|n\pi^0\right\rangle = \left|I_1 = 1/2, I_2 = 1; -1/2, 0\right\rangle$$

$$= \sqrt{\frac{2}{3}}\left|3/2, -1/2\right\rangle + \sqrt{\frac{1}{3}}\left|1/2, -1/2\right\rangle .$$

This gives part (a). Part (b) $K \to \pi\pi$ is a bit tricky. Two pions can have $I = 2, 1, 0$, and only $I = 2$ is forbidden by the $|\Delta I| = 1/2$ rule. However, for $\pi^0\pi^0$ $I = 1$ is forbidden, because the relevant Clebsch–Gordan coefficient vanishes. See Table 1.2. It is natural to assume that only $I = 0$ is available to $\pi^+\pi^-$ as well, and that gives the prediction.

$$\left|\pi^+\pi^-\right\rangle = \left|I_1 = 1, I_2 = 1; +1, -1\right\rangle$$

$$= \sqrt{\frac{1}{6}}\left|2, 0\right\rangle + \sqrt{\frac{1}{2}}\left|1, 0\right\rangle + \sqrt{\frac{1}{3}}\left|0, 0\right\rangle ;$$

and

$$\left|\pi^-\pi^+\right\rangle = |I_1 = 1, I_2 = 1; -1, +1\rangle$$

$$= \sqrt{\frac{1}{6}}\,|2,0\rangle - \sqrt{\frac{1}{2}}\,|1,0\rangle + \sqrt{\frac{1}{3}}\,|0,0\rangle\,.$$

$$\left|\pi^0\pi^0\right\rangle = |I_1 = 1, I_2 = 1; 0, 0\rangle = \sqrt{\frac{2}{3}}\,|2,0\rangle - \sqrt{\frac{1}{3}}\,|0,0\rangle\,.$$

The coefficients for $I = 0$ are the same, but $\pi^+\pi^-$ has two ways to go, and $\pi^0\pi^0$ only one. The spurion is a cute idea, originally proposed by one of the founders of quantum mechanics and also one of my best professors in graduate school — Gregor Wentzel.

$$\left|p\pi^0\right\rangle = |I_1 = 1/2, I_2 = 1; +1/2, 0\rangle$$

$$= \sqrt{\frac{2}{3}}\,|3/2, +1/2\rangle - \sqrt{\frac{1}{3}}\,|1/2, +1/2\rangle\,;$$

$$\left|n\pi^+\right\rangle = |I_1 = 1/2, I_2 = 1; -1/2, +1\rangle$$

$$= \sqrt{\frac{1}{3}}\,|3/2, +1/2\rangle + \sqrt{\frac{2}{3}}\,|1/2, +1/2\rangle\,;$$

$$\left|n\pi^-\right\rangle = |I_1 = 1/2, I_2 = 1; -1/2, -1\rangle = |3/2, -3/2\rangle\,.$$

For the spurion, both $\pm 1/2$ are possible, but $m_1 + m_2$ has to match on both sides of the amplitude.

$$\left|sp + \Sigma^+\right\rangle = |I_1 = 1/2, I_2 = 1; +1/2, +1\rangle = |3/2, +3/2\rangle\,;$$

or

$$\left|sp + \Sigma^+\right\rangle = |I_1 = 1, I_2 = 1/2; -1/2, +1\rangle$$

$$= \sqrt{\frac{1}{3}}\,|3/2, +1/2\rangle + \sqrt{\frac{2}{3}}\,|1/2, +1/2\rangle\,.$$

$$\left|sp + \Sigma^-\right\rangle = |I_1 = 1/2, I_2 = 1; -1.2, -1\rangle = |3/2, -3/2\rangle\,;$$

or

$$\left|sp + \Sigma^-\right\rangle = |I_1 = 1/2, I_2 = 1; +1/2, -1\rangle$$

$$= \sqrt{\frac{1}{3}}\,|3/2, -1/2\rangle - \sqrt{\frac{2}{3}}\,|1/2, -1/2\rangle\,.$$

Matching the magnetic quantum numbers gives:

$$\langle sp + \Sigma^+ | H_{3/2} + H_{1/2} | p\pi^0 \rangle = A_0 = \frac{\sqrt{2}}{3}(A_3 - A_1);$$

$$\langle sp + \Sigma^+ | H_{3/2} + H_{1/2} | n\pi^+ \rangle = A_+ = \frac{1}{3}A_3 + \frac{2}{3}A_1;$$

and

$$\langle sp + \Sigma^- | H_{3/2} + H_{1/2} | n\pi^- \rangle = A_- = A_3.$$

So $\sqrt{2}A_0 = A_- - A_+$, or $\sqrt{2}A_0 + A_+ - A_- = 0$. Since these are complex numbers, they form a closed triangle in a two-dimensional complex space. From the Particle Data Group we have:

$$\Sigma^+ \quad 0.8 \times 10^{-10} \text{ sec lifetime}$$
$$p\pi^0 \quad 51\% \text{ branching fraction}$$
$$n\pi^+ \quad 48\% \text{ branching fraction}$$
$$\Sigma^- \quad 1.5 \times 10^{-10} \text{ sec lifetime}$$
$$n\pi^- \quad 100\% \text{ branching fraction}$$

The total Σ^+ decay rate is twice the Σ^-, so all three of the decay amplitudes have the same magnitude, and $\sqrt{2}A_0$ just fits on an isosceles right triangle. There is more symmetry to the problem, exhibited in Problem 6.14. The amplitudes are real, but there are two of them because parity is not conserved. A_+ is pure P wave, $l = 1$, and A_- is pure S wave, $l = 0$. The two amplitudes are equal for A_0, giving maximum parity violation in the decay $\Sigma^+ \to p\pi^0$.

Problem 1.11. This is a numerical problem. Take Equation (1.26) for the $\bar{K}_0$ content of a beam that starts out pure K_0. Measure time in the K_0 rest frame in units of the K_s lifetime $\tau_s = 0.89510^{-10}$ s. Show that the mass difference term $\Delta mt/\tau_s = 0.47t/\tau_s$, and for $t/\tau_s = 0$ to 10, $t/\tau_L \approx 0$, so that we may write:

$$|\langle \bar{K}_0 | K_0 \rangle|^2 = (e^{-t} + 1 - 2\cos(0.47t)e^{-t/2})/4; t \text{ in units of } \tau_s.$$

$$(1.5)$$

Plot this function for $t/s = 0$ to 10. It flattens out at $1/4$ for large times. Why?

Solution 1.11. The equation of interest is Equation (1.38), not (1.26). It reads:

$$|\langle \bar{K}_0 | K_0 \rangle|^2 = (e^{-t/\tau_s} + e^{-t/\tau_L} - 2\cos(\Delta m t)e^{-(1/2\tau_s + 1/2\tau_L)t})/4.$$

From the Particle Data Group:

$$\tau_s = 0.895 \times 10^{-10} \text{ sec}; \quad \tau_L = 5.11 \times 10^{-8} \text{ sec};$$

$$\Delta m = m_L - m_s = 0.529 \times 10^{10} \hbar \sec^{-1}.$$

So $\Delta m \tau_s = 0.473$ as stated in the problem, and $\tau_L/\tau_s = 550$, so at $t = 10\tau_s$ it is safe to set $t/\tau_L = 0$. This gives Equation (1.5), where time is measured in units of τ_s. This function is plotted in Figure 1. The same curve, with the x-axis in distance instead of proper time, is shown in Figure 9.9 in the text. The probability reaches 1/4 for long times because we have ignored the K_L decay, so the beam is pure K_L, and we began with pure K^0, which is 50% K_L, and K_L is 50% $\bar{K}^0$.

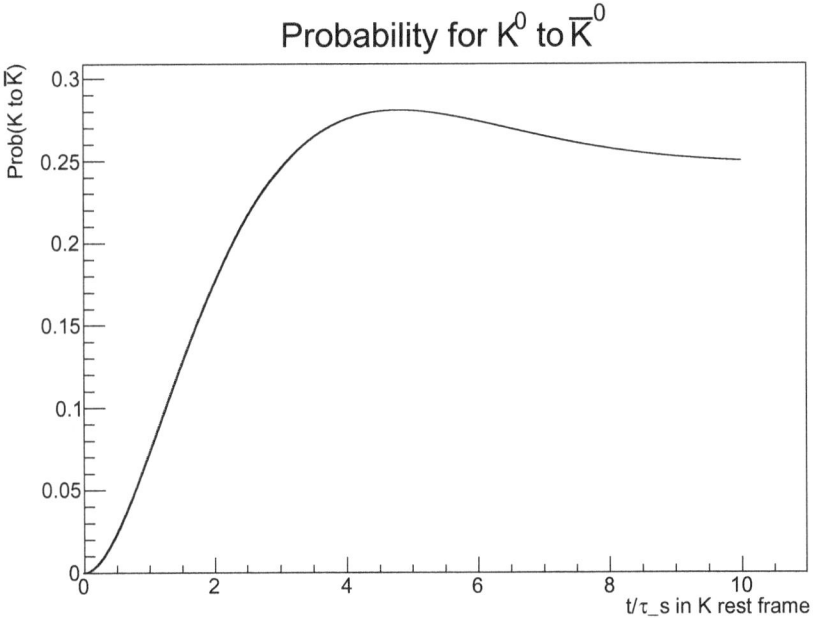

Figure 1: Probability that $K^0 \to \bar{K}^0$ as a function of proper time in units of $\tau_s = 0.895 \times 10^{-10}$ sec.

Colliding Beams

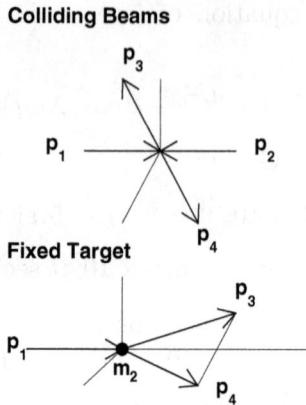

Figure 1.11: Kinematic diagrams for colliding beams and fixed target. In a single ring particle–antiparticle collider the total momentum is zero, and the energy available is twice the ring energy: $\sqrt{s} = 2E$. For a fixed target arrangement $\vec{p_2} = 0$, and $s = m_1^2 + m_2^2 + 2m_2E_1$. The two final state momenta share a plane with the incident beam direction.

Problem 1.12. Kinematics. Relativity is covered in Chapter 2, but we have already seen some relativistic formulas: s, t, and u. These quantities are invariants, meaning that they are the same in any rest frame. Figure 1.11 shows two typical experimental configurations for studying $2 \rightarrow 2$ reactions. Colliding beams are above, and fixed target is below. In a single ring collider, with counter rotating particle and antiparticle beams, the two energies must be equal, $E_1 = E_2$, and the total momentum must be zero, $\vec{p_2} = -\vec{p_1}$, so $s = 4E^2$. Now calculate s for the fixed target arrangement, where $E_2 = m_2$, and $\vec{p_2} = 0$. Show that in the high energy limit, where all masses can be ignored, $s = 2E_1m_2$. The energy available in the collision is $\sqrt{s}$, and $\sqrt{s} = 2E$ for a collider, while $\sqrt{s} = \sqrt{2E_1m_2}$ for fixed target. That is one reason why colliders are so popular.

Calculate the momentum transfer $t = (p_1 - p_3)^2$. Show that if the masses are neglected, $t = -4E^2 \sin^2 \theta/2$. Hence $1/t^2 = 1/(16E^4 \sin^4 \theta/2)$, which is the angle dependence of Rutherford scattering.

Solution 1.12. Collider with equal and opposite momenta; $s = (p_1+p_2)^2 = (E_1+E_2)^2 = 4E^2$, where E is the energy of the ring, and for a fixed target configuration $s = (p_1 + p_2)^2 = (E_1 + m_2)^2 - |\vec{p}_1|^2 = m_1^2 + m_2^2 + 2E_1m_2$. The four momentum transfer $t = (p_1 - p_3)^2 = m_1^2 + m_3^2 - 2p_1 \cdot p_3) = m_1^2 + m_3^2 - 2E_1E_3 + 2\vec{p}_1 \cdot \vec{p}_3$. In the high energy limit in the center of mass the energies are equal, so $t = -2E^2(1 - \cos\theta) = -4E^2\sin^2\theta/2$.

Problem 1.13. Show that $s + t + u = m_1^2 + m_2^2 + m_3^2 + m_4^2$.

Solution 1.13. $s + t + u = (p_1 + p_2)^2 + (p_1 - p_3)^2 + (p_1 - p_4)^2$ with $p_1 + p_2 = p_3 + p_4$, and $p_1^2 = m_1^2$ etc.

$$s + t + u = m_1^2 + m_2^2 + m_3^2 + m_4^2 + 2m_1^2 + 2p_1 \cdot p_2 - 2p_1 \cdot (p_3 + p_4)$$
$$= m_1^2 + m_2^2 + m_3^2 + m_4^2;$$

because $2m_1^2 + 2p_1 \cdot p_2 - 2p_1 \cdot (p_1 + p_2) = 0$.

Problem 1.14. Use the formula for s for fixed target to calculate the threshold energy for incident pions in $\pi^- + p \to \Lambda + K^0$. The threshold is defined as the kinetic energy that must be supplied to make a reaction proceed. At threshold $s = (m_3 + m_4)^2$. $m_3 + m_4$ is the minimum energy in the center of mass, but remember that s is an invariant, and can be calculated in any frame.

Solution 1.14. $s = m_\pi^2 + m_p^2 + 2m_pE_\pi = (m_\Lambda + m_K)^2$. Numbers: $m_\pi = 140\,\mathrm{MeV}$; $m_p = 938\,\mathrm{MeV}$; $m_K = 498\,\mathrm{MeV}$; $m_\Lambda = 1115\,\mathrm{MeV}$.

$$E_\pi = \frac{(m_\Lambda + m_K)^2 - m_\pi^2 - m_p^2}{2m_p} = 907\,\mathrm{MeV}.$$

Chapter 2

Problem 2.1. Show that in the expression for a rotation through three Euler angles about sequential axes:
$$R(\alpha, \beta, \gamma) = e^{-i\gamma L_{z''}} \times e^{-i\beta L_{y'}} \times e^{-i\alpha L_z} \qquad (2.210)$$
can be written in terms of fixed axes if the order is reversed:
$$R(\alpha, \beta, \gamma) = e^{-i\alpha L_z} \times e^{-i\beta L_y} \times e^{-i\gamma L_z}. \qquad (2.211)$$
Thus, you can erase the primes if you invert the order. The simplest way to do this is use $e^{-i\gamma L_{z''}} = e^{-i\beta L_{y'}} \times e^{-i\gamma L_z} \times e^{i\beta L_{y'}}$, and $L_{z'} = L_z$.

Solution 2.1. Follow the suggestion:
$$R(\alpha, \beta, \gamma) = e^{-i\gamma L_{z''}} \times e^{-i\beta L_{y'}} \times e^{-i\alpha L_z} = e^{-i\beta L_{y'}} \times e^{-i\gamma L_z} \times e^{-i\alpha L_z}$$
$$= e^{-i\alpha L_z} \times e^{-i\beta L_y} \times e^{-i\gamma L_z}.$$
Remember that $e^{i\alpha L_z}$ and $e^{-i\gamma L_z}$ commute.

Problem 2.2. The simplest rotation matrices are for spin $1/2$. Because $\vec{S} = \vec{\sigma}/2$, the rotation matrices involve $1/2$ angles. Show that:
$$e^{-i\alpha S_z} = \cos(\alpha/2) - i\sigma_z \sin(\alpha/2). \qquad (2.212)$$
This formula holds for all three components. Obtain the rotation matrix for Euler angles α and β:
$$R(\alpha, \beta) = e^{-i\alpha\sigma_z/2} \times e^{-i\beta\sigma_y/2}$$
$$= \begin{pmatrix} e^{-i\alpha/2}\cos(\beta/2) & -e^{-i\alpha/2}\sin(\beta/2) \\ e^{i\alpha/2}\sin(\beta/2) & e^{i\alpha/2}\cos(\beta/2) \end{pmatrix}. \qquad (2.213)$$

Since $e^{i\pi} = -1$, this operation gives a minus sign for rotation about the z axis through $360°$. The two component spin function χ transforms as $R(2\pi)\chi = -\chi$.

Solution 2.2. Expand the exponent, use $\sigma_z^2 = 1$, and collect terms:

$$e^{-i\alpha\sigma_z/2} = 1 - \frac{1}{2!}\left(\frac{\alpha}{2}\right)^2 + \frac{1}{4!}\left(\frac{\alpha}{2}\right)^4 - \cdots - i\left(\frac{\alpha}{2} - \frac{1}{3!}\left(\frac{\alpha}{2}\right)^3\right.$$
$$\left. + \frac{1}{5!}\left(\frac{\alpha}{2}\right)^5 - \cdots\right)\sigma_z;$$

Then the product

$$R(\alpha, \beta) = \cos(\alpha/2)\cos(\beta/2) - i(\cos(\alpha/2)\sin(\beta/2)\sigma_y$$
$$+ \sin(\alpha/2)\cos(\beta/2)\sigma_z - \sin(\alpha/2)\sin(\beta/2)\sigma_x);$$

where we have used $\sigma_y\sigma_z = i\sigma_x$. Then using the unit matrix, and the standard forms for σ_x, σ_y, and σ_z gives the result.

Problem 2.3. The next rotation matrices are for $l = 1$. Set the Euler angle $\gamma = 0$. For the Cartesian coordinates (x, y, z) the 3×3 rotation matrix is

$$\begin{pmatrix} x'' \\ y'' \\ z'' \end{pmatrix} = \begin{pmatrix} \cos\beta\cos\alpha & \cos\beta\sin\alpha & -\sin\beta \\ -\sin\alpha & \cos\alpha & 0 \\ \sin\beta\cos\alpha & \sin\beta\sin\alpha & \cos\beta \end{pmatrix} \begin{pmatrix} x \\ y \\ z \end{pmatrix}. \qquad (2.214)$$

The usual polar coordinates are

$$x = r\sin\theta\cos\phi, \qquad (2.215)$$
$$y = r\sin\theta\sin\phi, \qquad (2.216)$$
$$z = r\cos\theta. \qquad (2.217)$$

The spherical harmonics for $l = 1$ are

$$Y_1^{+1} = -(3/(8\pi))^{1/2}\sin\theta e^{i\phi}, \qquad (2.218)$$
$$Y_1^0 = (3/(4\pi))^{1/2}\cos\theta, \qquad (2.219)$$
$$Y_1^{-1} = (3/(8\pi))^{1/2}\sin\theta e^{-i\phi}. \qquad (2.220)$$

Obtain the mixing matrix that transforms the three spherical harmonics into Cartesian coordinates:

$$x = (4\pi/3)^{1/2} r M Y. \tag{2.221}$$

Show that M is unitary: $M^{-1} = M^\dagger$; and derive the $l = 1$ rotation matrix elements from $M^\dagger R M$, where R is the Cartesian matrix above.

Solution 2.3. The rotation matrix for (x, y, z) is the product of two matrices, using the "moving" axes, that is:

$$R(\alpha, \beta) = \begin{pmatrix} \cos\beta & 0 & -\sin\beta \\ 0 & 1 & 0 \\ \sin\beta & 0 & \cos\beta \end{pmatrix} \times \begin{pmatrix} \cos\alpha & \sin\alpha & 0 \\ -\sin\alpha & \cos\alpha & 0 \\ 0 & 0 & 1 \end{pmatrix}.$$

The matrix M is:

$$M = \begin{pmatrix} -1/\sqrt{2} & 0 & 1/\sqrt{2} \\ i/\sqrt{2} & 0 & i/\sqrt{2} \\ 0 & 1 & 0 \end{pmatrix}$$

The rotation matrix for $l = 1$, rows labeled $m_l = +1, m_l = 0, m_l = -1$ is:

$$R^{l=1}(\alpha, \beta) = M^\dagger \times R \times M$$

$$= \begin{pmatrix} e^{-i\alpha}(1 + \cos\beta)/2 & \sin\beta/\sqrt{2} & e^{i\alpha}(1 - \cos\beta)/2 \\ -\sin\beta e^{-i\alpha}/\sqrt{2} & \cos\beta & \sin\beta e^{i\alpha}/\sqrt{2} \\ e^{-i\alpha}(1 - \cos\beta)/2 & -\sin\beta/\sqrt{2} & e^{i\alpha}(1 + \cos\beta)/2 \end{pmatrix};$$

$$R^\dagger = R^{-1}.$$

This is the transpose of Equation (2.224), the result of Problem 2.4.

Problem 2.4. Another way to obtain the rotation matrix elements is to use the operator definition $R(\alpha, \beta) = e^{-i\alpha L_z} \times e^{-i\beta L_y}$, and the form for $l = 1$ used in Problem 1.4 on G-parity:

$$e^{-i\beta L_y} = 1 + (\cos\beta - 1)L_y^2 - i\sin\beta L_y, \tag{2.222}$$

and

$$e^{-i\alpha L_z} = 1 + (\cos\alpha - 1)L_z^2 - i\sin\alpha L_z. \tag{2.223}$$

The second equation is easy to handle because both L_z and L_z^2 are diagonal. Multiply the matrices together to give the matrix: $D^1_{m',m}(\alpha, \beta)$, which is the transpose of the rotation matrix of Problem 2.3, namely,

$$D^1_{m',m}(\alpha, \beta)$$

$$= \begin{pmatrix} e^{-i\alpha}(1+\cos\beta)/2 & -e^{-i\alpha}\sin\beta/\sqrt{2} & e^{-i\alpha}(1-\cos\beta)/2 \\ \sin\beta/\sqrt{2} & \cos\beta & -\sin\beta/\sqrt{2} \\ e^{i\alpha}(1-\cos\beta)/2 & e^{i\alpha}\sin\beta/\sqrt{2} & e^{i\alpha}(1+\cos\beta)/2 \end{pmatrix}.$$

$$(2.224)$$

The matrices are transposed because of the convention for the D matrices:

$$Y_1^m(\theta', \phi') = \sum_{m'=-1}^{+1} Y_1^{m'}(\theta, \phi) D^1_{m',m}(\alpha, \beta, \gamma). \tag{2.225}$$

Solution 2.4. The matrix L_y for $l = 1$ is given in Problem 1.4:

$$L_y = \begin{pmatrix} 0 & -i/\sqrt{2} & 0 \\ i/\sqrt{2} & 0 & -i/\sqrt{2} \\ 0 & i/\sqrt{2} & 0 \end{pmatrix} ; \quad L_y^2 = \begin{pmatrix} 1/2 & 0 & -1/2 \\ 0 & 1 & 0 \\ -1/2 & 0 & 1/2 \end{pmatrix} ;$$

$$e^{-i\beta L_y} = \begin{pmatrix} (1+\cos\beta)/2 & -\sin\beta/\sqrt{2} & (1-\cos\beta)/2 \\ \sin\beta/\sqrt{2} & \cos\beta & -\sin\beta/\sqrt{2} \\ (1-\cos\beta)/2 & \sin\beta/\sqrt{2} & (1+\cos\beta)/2 \end{pmatrix} ;$$

$$e^{-i\alpha L_z} = \begin{pmatrix} e^{-i\alpha} & 0 & 0 \\ 0 & 0 & 0 \\ 0 & 0 & e^{i\alpha} \end{pmatrix}.$$

The product of these two matrices is Equation (2.224).

Problem 2.5. Calculate the determinant of the 'proper' Lorentz transformation Equation (2.25). Write down the Lorentz transformation for time reversal $t \to -t$ and no other changes. Then write down the Lorentz transformation for space inversion $\vec{x} \to -\vec{x}$ and no other changes. Calculate these determinants. They are called 'improper' Lorentz transformations.

Solution 2.5. Determinants of Lorentz transformations. Proper one:

$$\begin{bmatrix} \gamma & 0 & 0 & -v\gamma \\ 0 & 1 & 0 & 0 \\ 0 & 0 & 1 & 0 \\ -v\gamma & 0 & 0 & \gamma \end{bmatrix} = \gamma \times \gamma + v\gamma \times (-v\gamma) = \gamma^2(1 - v^2) = 1;$$

where we have evaluated the determinant using minors and cofactors. Parity and time reversal are easy, because there is only one term:

$$\begin{bmatrix} 1 & 0 & 0 & 0 \\ 0 & -1 & 0 & 0 \\ 0 & 0 & -1 & 0 \\ 0 & 0 & 0 & -1 \end{bmatrix} = -1; \qquad \begin{bmatrix} -1 & 0 & 0 & 0 \\ 0 & 1 & 0 & 0 \\ 0 & 0 & 1 & 0 \\ 0 & 0 & 0 & 1 \end{bmatrix} = -1.$$

Problem 2.6. Rapidity. The Lorentz transformation parameters $\vec{v}$ and $\gamma = 1/\sqrt{1 - v^2}$ are related to hyperbolic functions. Show that in terms of the angle χ, called the rapidity, one can use the trigonometric identities for hyperbolic functions to write:

$$v = \tanh \chi, \qquad (2.226)$$

$$\gamma = \cosh \chi, \qquad (2.227)$$

$$v\gamma = \sinh \chi. \qquad (2.228)$$

Show that the hyperbolic functions have the correct ranges: $1 \leq \gamma \leq \infty$ and $-1 \leq v \leq +1$. While in relativity velocities do not simply add, rapidities do. For two collinear Lorentz transformations, $\chi = \chi_1 + \chi_2$. Show that Einstein's velocity addition theorem $v = (v_1 + v_2)/(1 + v_1 v_2)$ follows from the rules for combining hyperbolic functions. Express the Lorentz transformation in terms of χ.

Solution 2.6. Rapidity is a very useful concept. By definition:

$$\sinh \chi = (e^\chi - e^{-\chi})/2; \quad \cosh \chi = (e^\chi + e^{-\chi})/2;$$

$$\tanh \chi = \sinh \chi / \cosh \chi; \quad \cosh^2 \chi - \sinh^2 \chi = 1;$$

and

$$v = \tanh \chi; \quad \gamma = \cosh \chi; \quad v\gamma = \sinh \chi; \text{ Lorentz transformation}$$

$$L = \begin{pmatrix} \cosh \chi & 0 & 0 & \sinh \chi \\ 0 & 1 & 0 & 0 \\ 0 & 0 & 1 & 0 \\ \sinh \chi & 0 & 0 & \cosh \chi \end{pmatrix}.$$

The Lorentz transformation in terms of rapidity resembles a two-dimensional rotation, with two differences: the functions of the "angle" rapidity are hyperbolic, and the minus sign is absent on the $\sinh \chi$ term in the lower left corner.

Problem 2.7. More rapidity. Lorentz transformations by convention are normally along the z axis. Momentum components perpendicular to the z axis, called $p_\perp$, are Lorentz invariant, while parallel components, $p_\parallel$ change from one moving frame to the next. It is convenient to introduce a variation of the rapidity χ, called y, defined as

$$y = (1/2) \ln\big((E + p_\parallel)/(E - p_\parallel)\big). \tag{2.229}$$

While it may seem confusing, this definition is also called rapidity. Show that

$$\tanh y = p_\parallel/E, \tag{2.230}$$

$$\cosh y = E/\sqrt{E^2 - p_\parallel^2}. \tag{2.231}$$

Since $E^2 = m^2 + p_\perp^2 + p_\parallel^2$, the denominator in $\cosh y$ is often called the 'transverse mass'. Like the mass of a particle, the transverse mass is Lorentz invariant. Rapidities add, and differences in rapidity — rapidity gaps — are Lorentz invariant. If the total energy is $\sqrt{s}$, the maximum energy of a pion is $\sqrt{s}/2$, because there has to be another pion to balance the momentum. The maximum value of $p_\parallel = \pm\sqrt{s/4 - m^2}$. Show that for $\sqrt{s} \gg m$, $y_{max} = \pm \ln(\sqrt{s}/m)$. Calculate the full spread of rapidity for pions with $\sqrt{s} = 2\,\text{TeV}$.

Solution 2.7. From the definition:

$$e^y = \sqrt{\frac{E + p_{\parallel}}{E - p_{\parallel}}}; \quad e^{-y} = \sqrt{\frac{E - p_{\parallel}}{E + p_{\parallel}}}.$$

Results for the hyperbolic functions follow from these two formulas. The transverse mass defined by $m_{\perp}^2 = m^2 + p_{\perp}^2$ is always greater than the true mass. The same term "transverse mass" is also applied to Equation (7.47), the observable used to measure the mass of the W boson, but Equation (7.47) is always less than the true mass of the W.

Problem 2.8. Still more rapidity. Pseudorapidity $\eta = (1/2) \ln((1 + \cos\theta)/(1 - \cos\theta)) = -\ln(\tan(\theta/2))$ is y for a massless particle. η diverges as $\theta \to 0$. Detectors are often segmented in $|\Delta\eta|$, because this results in constant particle flux per segment. How many towers would cover $\theta = 90°$ to $\theta = 20°$ in steps $\Delta\eta = 0.1$? The towers are also segmented in $\Delta\phi$ in the transverse plain.

Solution 2.8. Use the inverse formula for the polar angle θ:

$$\tan(\theta/2) = e^{-\eta}; \quad \tan\theta = (2e^{-\eta})/(1 - e^{-2\eta}).$$

Then start at $\theta = 90°, \eta = 0$ and march forward in steps of $\Delta\eta = 0.1$. Near $\theta = 90°$ the $\Delta\theta$ steps are about $5.5°$, while near $\theta = 20°$ they are about $2.5°$. I count 18 towers for $20° \leq \theta \leq 90°$. The segmentation is shown in Figure 1. The cover of the book shows one of the central calorimeter arches retracted from the detector on the right-hand side of the photograph. There are four such arches, each covering $180°$ in 12 ϕ segments. The $\Delta\eta = 0.1$ segmentation covers the polar angle region from $90°$ to about $45°$ in 10 bins.

Problem 2.9. Two-body decay kinematics. A two-body decay like $K_s^0 \to \pi^+ + \pi^-$, $K^+ \to \mu^+ + \nu_\mu$, or $\pi^0 \to \gamma + \gamma$ is completely determined in the parent rest frame by the masses $m \to m_1 + m_2$. The momenta are equal and opposite, and the energies are given by Equation (2.63): $E_1 = (m^2 + m_1^2 - m_2^2)/(2m)$. The same formula with m_1 and m_2 interchanged gives E_2. If $m_1 = m_2$ then $E_1 = E_2 = m/2$. The momenta can be calculated from the energies, or from Equation (2.64). If the parent has no spin (like π's and K's),

η **Segmentation**

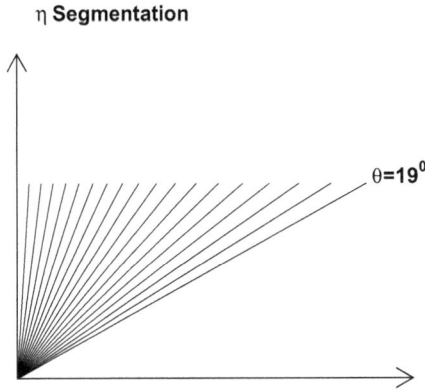

Figure 1: Segmentation in polar angle of the central electromagnetic and hadronic calorimeters in CDF. The central calorimeter arches are shown retracted in the figure on the cover of the book. Azimuthal angle segmentation $\Delta\phi = 15^0$ is clearly visible, but the polar angle segmentation is hidden from view.

the distribution of momentum vectors $\vec{p_1}$ is isotropic in the parent rest frame, which is flat in $\cos\theta_c$. Show that an isotropic angular distribution in the parent rest frame maps into a uniform distribution in energy in the laboratory. Calculate the maximum and minimum energies for particle #1 in the laboratory — the ends of the box distribution.

Solution 2.9. The energy of particle one in the laboratory is $E_1^{lab} = \gamma(E_1 + vp_1\cos\theta)$. $dE_1^{lab}/d\cos\theta = \gamma v p_1$, a constant. θ is the polar angle in the parent particle rest frame.

Problem 2.10. If the parent is moving in the laboratory, $\vec{p_1}$ and $\vec{p_2} = -\vec{p_1}$ must be Lorentz transformed to give the final state momenta. Equation (2.62) gives the polar angle transformation formula for particle #1. The azimuthal angle is Lorentz invariant. For particle #2 $\cos\theta_c \to -\cos\theta_c$, and if $m_2 \neq m_1$ the parameter $a = v_c E_2/p_2$ has to be adjusted accordingly. One parameter of experimental interest is the opening angle between the two final state particles in the lab frame. Use the trigonometric formula for adding tangents to obtain the expression for $m_1 = m_2$:

$$\tan(\theta_1 + \theta_2) = (2a\gamma\sin\theta_c)/(\gamma^2(a^2 - \cos^2\theta_c) - \sin^2\theta_c). \quad (2.232)$$

Show that this formula has two possible extremes $d\tan(\theta_1 + \theta_2)/d\theta_c = 0$:

$$\cos\theta_c = 0, \quad \text{or} \quad \sin^2\theta_c = (\gamma^2(a^2 - 1))/(\gamma^2 - 1). \quad (2.233)$$

The second condition can be satisfied if $a > 1$. This can be true for $K \rightarrow \pi\pi$, but not for $\pi^0 \rightarrow \gamma + \gamma$.

Solution 2.10. Equation (2.62) for an equal mass two-body decay reads:

$$\tan\theta_1 = \sin\theta_c/(\gamma(\cos\theta_c + a)); \quad \text{and} \quad \tan\theta_2 = \sin\theta_c/\gamma(-\cos\theta_c + a)).$$

Use the trig identity:

$$\tan(\theta_1 + \theta_2) = (\tan\theta_1 + \tan\theta_2)/(1 - \tan\theta_1\tan\theta_2);$$

to get Equation (2.232):

$$\tan(\theta_1 + \theta_2) = (2a\gamma\sin\theta_c)/(\gamma^2(a^2 - \cos^2\theta_c) - \sin^2\theta_c).$$

The derivative is:

$$\frac{d\tan(\theta_1 + \theta_2)}{d\theta_c} = \frac{2a\gamma\cos\theta_c}{(\gamma^2(a^2 - \cos^2\theta_c) - \sin^2\theta_c)^2}$$
$$\times (\gamma^2(a^2 - 1) + (1 - \gamma^2)\sin^2\theta_c).$$

This expression vanishes for either of the conditions in Equation (2.233). The second condition, $\sin^2\theta_c = (\gamma^2(a^2 - 1))/(\gamma^2 - 1)$, can only be satisfied if $a > 1$, or the velocity of the parent is greater than the velocity of either daughter in the center of mass. This condition therefore depends on the energy of the parent in the laboratory, provided that the mass of the daughters is greater than zero. For $K \rightarrow \pi\pi$, $a = 1$ at about 880 MeV total K energy.

Problem 2.11. $\pi^0 \rightarrow \gamma + \gamma$. Here $a < 1$ because the π^0 velocity is always less than the γ. Calculate the energy of the γ emitted backwards in the laboratory for a 1.35 GeV π^0 ($\gamma = 10$). Plot the opening angle distribution vs. $\cos\theta_c$ for γ's from 1.35 GeV π^0's. Although not strictly correct, you may set $a = 1$ in the angle formula. Rather than a maximum value — ($\theta_1 + \theta_2 = \pi$ is always possible), the opening angle distribution has a minimum value $\theta_1 + \theta_2 \approx 2/\gamma$. The opening angle distribution of pairs of γ's is sharply peaked near the minimum — the γ's lie on a cone. A simple way to obtain the minimum opening angle is to look at a pair of γ's at $\theta_c = \pi/2$. Invariant transverse momentum $p_\perp = m/2$ and in the lab frame $p_\parallel = \gamma m/2$, so the opening angle is $2/\gamma$.

Solution 2.11. High energy $\pi^0 \to \gamma + \gamma$. Here $a < 1$ always, because $v = 1$ for the $\gamma's$. For $\gamma = 10$ in the backward direction in the lab we have:

$$E_\gamma = \gamma(1 - v)m_\pi/2 = m_\pi/40 = 3.4\,\text{MeV}.$$

$v = a = 0.995$, and $v^2 = a^2 = 0.99$. The formula for $\tan(\theta_1 + \theta_2)$ simplifies if we set $v = 1$:

$$\tan(\theta_1 + \theta_2) = \frac{2v\gamma \sin\theta_c}{\gamma^2(v^2 - \cos^2\theta_c) - \sin^2\theta_c} \sim \frac{2\gamma}{(\gamma^2 - 1)\sin\theta_c} = \frac{.2}{\sin\theta_c}.$$

This function is plotted in Figure 2.

Problem 2.12. Calculate v_c for either pion in the decay $K_s^0 \to \pi^+ + \pi^-$. The parameter a can be any number $0 \le a \le 1/v_c$. For K_0 beam momentum of 500 MeV/c plot a curve of $\theta_1 + \theta_2$ vs. $\cos\theta_c$. The opening angle distribution should have a minimum value (about $79°$, coming from $\theta_c = \pi/2$). Repeat the calcultaion of the opening angle distribution for 1500 MeV/c K_0's. In this case the distribution has a maximum opening angle of about $33°$ that does not occur for $\theta_c = \pi/2$, but rather for an asymmetric two pion decay in the K rest frame. Even simple two-body kinematics has some surprises!

Solution 2.12. $K_s^0 \to \pi^+\pi^-$ angle transformation. Numbers: $m_K = 498$ MeV, $m_\pi = 140$ MeV. In the K rest frame $E_\pi = M_K/2 = 249$ MeV; $p_\pi = 206$ MeV. For $p_K = 500$ MeV, $\gamma_K = 1.417$ $a = 0.856$; and for $p_K = 1500$ MeV $\gamma_K = 3.17$ $a = 1.476$. So in the first case $a < 1$, and the tangent function, shown in Figure 3, has a minimum opening angle of about 78^0, and resembles $\pi^0 \to \gamma + \gamma$. However, for the second case $a > 1$, and as Figure 4 shows, the opening angle distribution is very different. The opening angle distribution now has a maximum value rather than a minimum, and that maximum is no

$$\pi^0 \text{->} \gamma\gamma$$

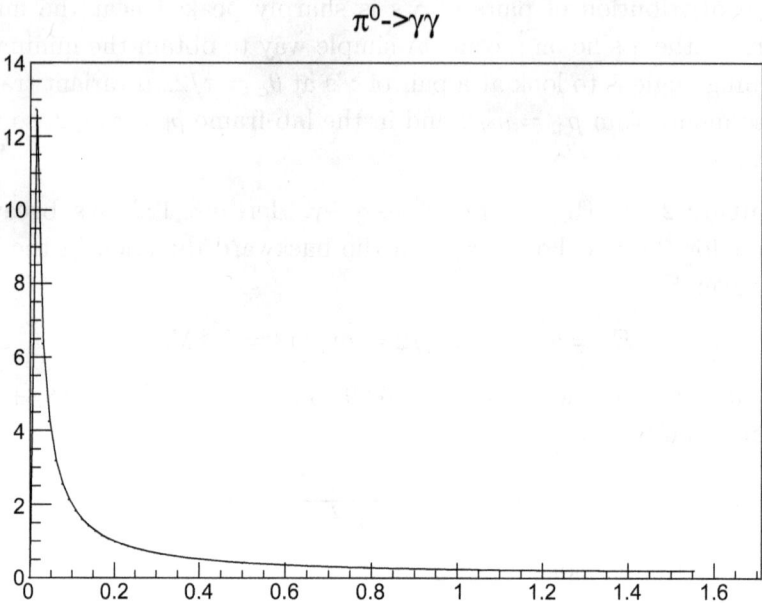

Figure 2: $\tan(\theta_1 + \theta_2)$ vs θ_1 in radians. The function used is singular at $\theta_1 = 0$, which is not correct; at $\theta_1 = 0$ the sum $\theta_1 + \theta_2 = \pi$, and $\tan \pi = 0$. However as shown the opening angle quickly moves to less than $\pi/2$, and most of the γ pairs are in a cone of opening angle $\theta_1 + \theta_2 \sim 0.2 = 11^0$ for a 1.35 GeV π^0.

longer at 90^0 in the K^0 rest frame. The maximum opening angle for the pion pair from 1500 MeV K^0's is about 34^0 with the forward pion at about 30^0 in the K^0 rest frame.

Problem 2.13. Three-body decays. An easy one to start with is $K^+ \rightarrow \pi^+\pi^+\pi^-$; all three masses are equal. Calculate the total kinetic energy available, the maximum energy of any pion, and the energy of pion #1 or #2 if #3 is at rest. The boundary curve lies within a square on a (x, y) plot, where $x = (E_1 - m_1)/(E_1^{\max} - m_1)$, and $y = (E_2 - m_2)/(E_2^{\max} - m_2)$; $0 \leq x \leq 1$, and $0 \leq y \leq 1$. The boundary curve is an oval inscribed in the square. Set $|\cos\theta_{12}| = 1$ in Equation (2.67), square both sides, and solve for the boundary curve $y = f(x)$.

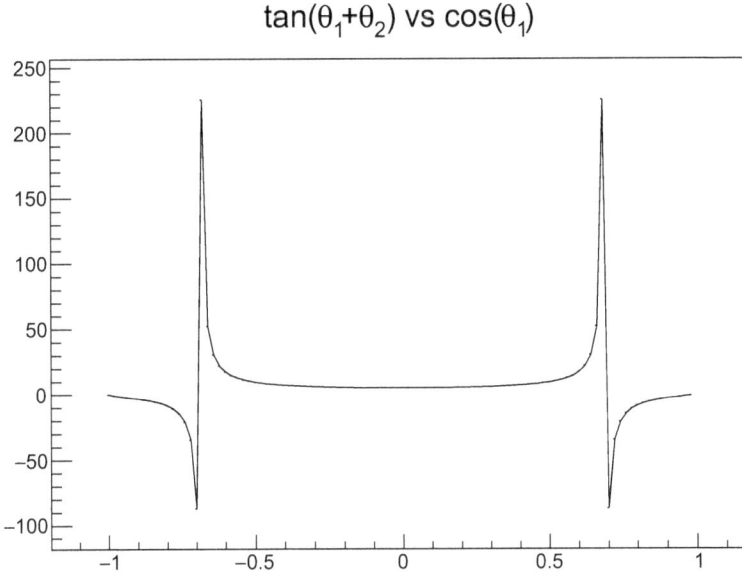

Figure 3: $\tan(\theta_1 + \theta_2)$ vs $\cos\theta_1$ for $K_s^0 \to \pi^+\pi^-$ at $p_K = 500\,\text{MeV}$. The pion velocity in the K rest frame is greater than the K velocity in the lab, so pions can appear backwards, and the maximum opening angle is 180^0. The tangent function diverges at 90^0, giving the peculiar wings to the plot. The minimum opening angle is about 78^0 for $\cos\theta_1 = 0$.

Solution 2.13. For $K \to 3\pi$ the boundary curve is:

$$AE_2^2 + BE_2 + C = 0; \quad A = 4(m_K^2 + m_\pi^2 - 2m_K E_1);$$
$$B = 4((3m_K^2 + m_\pi^2)E_1 - m_K(m_K^2 + m_\pi^2 + 2E_1^2));$$
$$C = m_K^4 + 2m_K^2 m_\pi^2 - 3m_\pi^4 - 4E_1(m_K^2 + m_\pi^2)(m_K - E_1).$$

Two pions do not have the maximum energy simultaneously. $E_1^{\min} = m_\pi$, and $E_1^{\max} = (m_K^2 - 3m_\pi^2)/(2m_K)$. As a check, set $E_1 = m_\pi$. Then

$$A = 4(m_K - m_\pi)^2; \quad B = -4(m_K - m_\pi)^3; \quad \text{and}$$
$$C = (m_K - m_\pi)^4; \text{ giving } E_2 = E_3 = \frac{m_K - m_\pi}{2}.$$

The pions are mildly relativistic. $E_1^{\max} = 188\,\text{MeV}$, so $\gamma = 1.34$ and $v = 0.66$.

$\tan(\theta_1+\theta_2)$ vs $\cos(\theta_1)$

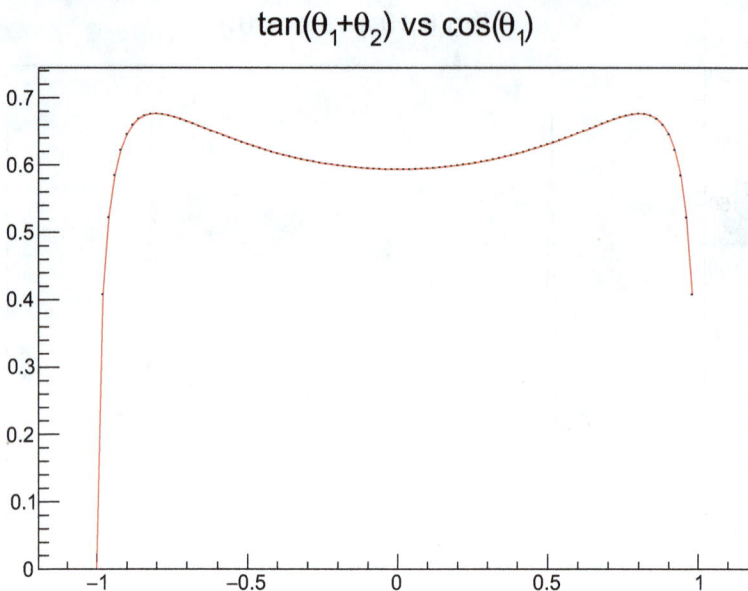

Figure 4: $\tan(\theta_1 + \theta_2)$ vs $\cos \theta_1$ for $K_s^0 \to \pi^+\pi^-$ at $p_K = 1500\,\text{MeV}$. Now the velocity of the K^0 in the lab exceeds the velocity of the pion in the K^0 rest frame: $a = 1.476$, and the opening angle distribution is very different. Pions cannot appear in the backward direction in the lab, and there is a maximum opening angle that comes from an asymmetric decay in the K^0 rest frame.

Problem 2.14. An alternate form of the kinematic plot is sometimes convenient if the three final state particles all have the same mass. The plot uses the fact that the sum of the three perpendiculars from any point inside an equilateral triangle is a constant. That constant is the total kinetic energy available, and the three perpendiculars are the kinetic energies of the particles. Show that the non-relativistic kinematic boundary is the inscribed circle. Obtain the boundary for $K \to \pi\pi\pi$. Show that the boundary curve for the rare decay $\rho^0 \to \gamma\gamma\gamma$ is an inverted inscribed equilateral triangle.

Solution 2.14. This is really a study of the geometry of an equilateral triangle. See Figure 5. Call the side a. Then the perpendiculars are $h = \sqrt{3}a/2$, and the radius (distance from the center to any side) is $r_0 = h/3$. Hence the total kinetic energy is $E_1 + E_2 + E_3 = E = h$.

This is the mass of the parent minus the masses of the three daughters, which are assumed to be all the same. We may think of $K \to 3\pi$, but we imagine that the kinetic energy is non-relativistic, or the K mass is only slightly greater than 3 times the pion mass. Using the non-relativistic kinetic energy formula together with the usual collinear requirement for momenta on the boundary leads to the analog of the equations in Problem 2.13, namely:

$$AE_2^2 + BE_2 + C = 0; \quad A = 4; \quad B = -4(E - E_1);$$
$$C = 4E_1^2 - 4EE_1 + E^2.$$

As a check, if $E_1 = 0$, then $E_2 = E_3 = E/2$; and $E_1 = E_1^{\max} = 2E/3$, then $E_2 = E_3 = E/6$. The boundary curve on the triangle is an inscribed circle. You can easily check that at the points where the curve intersects the main diagonals, that is $2r_0$ from the bases, each of the other two perpendiculars are $h/6$, conserving energy. As the decay becomes relativistic, the circle eventually collapses to an inscribed upside-down equilateral triangle. The base of this triangle intercepts the vertical perpendicular half way up at $h/2 = 3r_0/2$. Non-relativistic kinetic energy is proportional to momentum squared, and that leads to a circle. In the relativistic limit, energy and momentum are equal, and that leads to straight lines. The inverted triangle would apply to the annihilation of triplet positronium, discussed in Chapter 9.

Problem 2.15. $K^+ \to \mu^+ \pi^0 \nu_\mu$. Here one particle, the neutrino, has zero mass, and the muon and pion have maximum energy simultaneously. Calculate $E_\mu^{\max}$ and $E_\pi^{\max}$. Solve Equation (2.69) for the boundary curve, and plot it on an (x, y) plot, where $x = (E_\mu - m_\mu)/(E_\mu^{\max} - m_\mu)$, and $y = (E_\pi - m_\pi)/(E_\pi^{\max} - m_\pi)$. This reproduces Figure 2.2.

Problem 2.16. Repeat 2.15 for $K^+ \to e^+ \pi^0 \nu_e$. Here both the positron and the neutrino can be treated massless. This reproduces Figure 2.3.

Solution 2.15 and 2.16. This subject has been adequately covered. The reader should know what to do to plot the boundary curves for $K \to \pi\mu\nu$ and $K \to \pi e\nu$, Figures 2 and 3.

Problem 2.17. Relativistic wave equations. Show that the charge density ρ defined by Equation (2.91) reduces to $\rho = \psi^*\psi$ in the limit $E-m \ll m$, with the relativistic wave function $\phi(x,t) = \psi(x,t)e^{-imt}$. The non-relativistic wave function has the mass subtracted off from the energy.

Solution 2.17. Divide Equations (2.90) and (2.91) by m, so that 2.90 resembles the non-relativistic version of $\vec{j}$. This is because of different wave function normalizations. Then:

$$\rho = \frac{1}{2mi}\left(\frac{\partial\phi^*}{\partial t}\phi - \phi^*\frac{\partial\phi}{\partial t}\right);$$

$$\phi = \psi e^{-imt}; \quad \frac{\partial\phi}{\partial t} = \left(\frac{\partial\psi}{\partial t} - im\psi\right)e^{-imt};$$

and dropping terms $1/m$ gives $\rho \to \psi^*\psi$.

Problem 2.18. Show that the 4×4 Dirac matrices written in 2×2 boxes have all the proper characteristics:

$$\vec{\alpha} = \begin{pmatrix} 0 & \vec{\sigma} \\ \vec{\sigma} & 0 \end{pmatrix}, \quad \beta = \begin{pmatrix} 1 & 0 \\ 0 & -1 \end{pmatrix}. \tag{2.234}$$

Namely, $\alpha_i\alpha_j + \alpha_j\alpha_i = 2\delta_{i,j}$, $\beta\alpha_j + \alpha_j\beta = 0$, $\beta^2 = 1$, $(\vec{a}\cdot\vec{\alpha})(\vec{b}\cdot\vec{\alpha}) = \vec{a}\cdot\vec{b} + i\Sigma\cdot(\vec{a}\times\vec{b})$. The matrix $\vec{\Sigma}$ is defined in Equation (2.143).

Equilateral Triangle for K->3 π

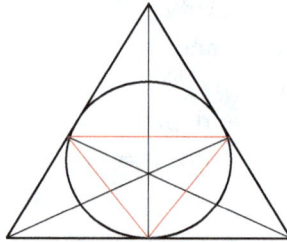

Figure 5:   Geometry of an equilateral triangle. It has been used as a phase space plot for three-body decays into equal mass particles, like $K \to \pi\pi\pi$, based on the property of the diagram that the sum of the three perpendiculars to the sides from any interior point is a constant. The length of a perpendicular is proportional to the kinetic energy of a particle. The kinematic boundary is the inscribed circle if the final state particles are non-relativistic. The curve is pushed in as the available energy increases, becoming the red triangle for three γ rays.

Solution 2.18. The operations are straightforward, using the Pauli spin properties $\sigma_x \sigma_y = i\sigma_z$, and $(\vec{a} \cdot \vec{\sigma})(\vec{b} \cdot \vec{\sigma}) = \vec{a} \cdot \vec{b} + i\vec{\sigma} \cdot (\vec{a} \times \vec{b})$.

Problem 2.19. Show that the spinors $u^{(s)}(p)$ with $s = 1, 2, 3, 4$ are consistent with Dirac's equation; $s = 1, 2$ for $E > 0$ and $s = 3, 4$ for $E < 0$. Check the normalizations and orthogonality for both $u^\dagger(p)u(p)$ and $\bar{u}(p)u(p)$.

Solution 2.19. The spinors are given in Equations (2.121), (2.122), and (2.129).

$$u^{(1)}(p) = \sqrt{E+m}\begin{pmatrix} 1 \\ 0 \\ \dfrac{p_z}{E+m} \\ \dfrac{p_x + ip_y}{E+m} \end{pmatrix} \; ; \quad u^{(2)}(p) = \sqrt{E+m}\begin{pmatrix} 0 \\ 1 \\ \dfrac{p_x - ip_y}{E+m} \\ \dfrac{-p_z}{E+m} \end{pmatrix} \; ;$$

$$\not{p} = \begin{pmatrix} E & 0 & -p_z & -p_x + ip_y \\ 0 & E & -p_x - ip_y & p_z \\ p_z & p_x - ip_y & -E & 0 \\ p_x + ip_y & -p_z & 0 & -E \end{pmatrix} \; ;$$

$$\not{p}\,u^{(1)}(p) = \sqrt{E+m}\begin{pmatrix} E - \dfrac{p^2}{E+m} \\ 0 \\ p_z - \dfrac{Ep_z}{E+m} \\ p_x + ip_y - E\dfrac{p_x + ip_y}{E+m} \end{pmatrix} = m u^{(1)}(p).$$

$u^{(2)}(p)$ is the same. For $u^{(3)}(p)$ you have to change the sign of E in $\not{p}$:

$$\not{p}\,u^{(3)}(p) = \sqrt{|E|+m}\begin{pmatrix} |E|\dfrac{p_z}{|E|+m} - p_z \\ |E|\dfrac{p_x + ip_y}{|E|+m} - p_x - ip_y \\ -\dfrac{p_z^2}{|E|+m} - \dfrac{p_x^2 + p_y^2}{|E|+m} + |E| \\ -\dfrac{p_z(p_x + ip_y)}{|E|+m} + \dfrac{p_z(p_x + ip_y)}{|E|+m} \end{pmatrix} = m u^{(3)}(p).$$

Problem 2.20. Diagonalize the Hamiltonian $H = \vec{\alpha} \cdot \vec{p} + \beta m$.

Solution 2.20. Let $\vec{p} = p\hat{z}$ for simplicity. Then we are faced with calculation of a 4×4 determinant, but there are lots of zeros. Thus:

$$
\begin{bmatrix}
m - \lambda & 0 & p & 0 \\
0 & m - \lambda & 0 & -p \\
p & 0 & -m - \lambda & 0 \\
0 & -p & 0 & -m - \lambda
\end{bmatrix} = 0;
$$

$$(m^2 - \lambda^2)^2 + 2(m^2 - \lambda^2)p^2 + p^4 = 0; \quad \lambda^2 = p^2 + m^2; \quad \lambda = \pm E.$$

We used the minors-cofactors rules to calculate the 4×4 determinant as four 3x3 determinants, two terms in this case being zero.

Problem 2.21. Check the algebra in reducing Equations (2.141)–(2.147).

Solution 2.21. The steps from Equation (2.141) to (2.147) are outlined in the text. For instance, Equation (2.146) goes like this:

$$(\vec{\alpha} \cdot (\vec{p} - e\vec{A}))(\beta m - (E - eV)) + (\beta m + (E - eV))(\vec{\alpha} \cdot (\vec{p} - e\vec{A}))$$
$$= -\vec{\alpha} \cdot (\vec{p} - e\vec{A})(E - eV) + (E - eV)(\vec{\alpha} \cdot (\vec{p} - e\vec{A})).$$

The mass terms cancel because $\vec{\alpha}\beta = -\beta\vec{\alpha}$. Now substitute $\vec{p} \to -i\vec{\nabla}$ and $E \to i\partial/\partial t$.

$$= -\vec{\alpha} \cdot \vec{\nabla}\frac{\partial}{\partial t} + ie\vec{\alpha} \cdot \vec{A}\frac{\partial}{\partial t} - ie\vec{\alpha} \cdot \vec{\nabla}V - e^2\vec{\alpha} \cdot \vec{A}V$$
$$+ \frac{\partial}{\partial t}\vec{\alpha} \cdot \vec{\nabla} - ie\frac{\partial}{\partial t}\vec{\alpha} \cdot \vec{A} + ieV\vec{\alpha} \cdot \vec{\nabla} + e^2V\vec{\alpha} \cdot \vec{A}.$$

Now operate on an arbitrary function $\psi(\vec{x}, t)$ to cancel terms, leaving:

$$= -ie\psi(\vec{x}, t)\vec{\alpha} \cdot (\vec{\nabla}V + \frac{\partial\vec{A}}{\partial t}) = ie\vec{\alpha} \cdot \vec{E};$$

where at the end we drop the arbitrary function.

Problem 2.22. Show that the last term in Equation (2.149), $(-ie\vec{\alpha} \cdot \vec{E})\psi/(2m)$ reduces to the spin–orbit force of Equation (2.148).

Solution 2.22. The correct form for Equation (2.152) is:

$$ie\vec{\sigma} \cdot \vec{E}/(2m)\psi_b \approx ((Ze^2)/(2m^2r^3)\vec{S} \cdot \vec{L})\psi_a.$$

In this approximation we have dropped a non-hermitian term $ie\vec{E} \cdot \vec{p}/(2m)^2\psi_a$. This is called the "Darwin term", and is discussed by Sakurai page 86 ff. For our discussion it is one more peculiar feature of approximations to the Dirac equation. The spin-orbit force however is correct.

$$-ie\frac{\vec{\alpha} \cdot \vec{E}}{2m}\psi = -ie\begin{pmatrix} 0 & \vec{\sigma} \cdot \vec{E}/(2m) \\ \vec{\sigma} \cdot \vec{E}/(2m) & 0 \end{pmatrix}\begin{pmatrix} \psi_a \\ \psi_b \end{pmatrix}.$$

The upper component is:

$$-ie\frac{\vec{\sigma} \cdot \vec{E}}{2m}\psi_b = -ie\frac{(\vec{\sigma} \cdot \vec{E})(\vec{\sigma} \cdot \vec{p})}{2m(E+m)}\psi_a.$$

replacing $E + m \approx 2m$, and using the $(\vec{a} \cdot \vec{\sigma})(\vec{b} \cdot \vec{\sigma})$ gives:

$$-ie\frac{\vec{\alpha} \cdot \vec{E}}{2m}\psi \approx -i\frac{e}{(2m)^2}(\vec{E} \cdot \vec{p} + i\vec{\sigma} \cdot (\vec{E} \times \vec{p}))\psi_a.$$

Substituting $\vec{E} = -\vec{\nabla}V = (Ze^2)/r^3\vec{r}$, and $\vec{S} = \vec{\sigma}/2$ gives the spin-orbit term.

Problem 2.23. Verify the completeness relations Equations (2.165) and (2.166) by multiplying from the right by $u^{(r)}(p)$ or $v^{(r)}(p)$, using the normalization, and the Dirac equation.

Solution 2.23. Completeness sums, Equations (2.165) and (2.166)

$$\sum_{s=1,2} u^{(s)}(p)\bar{u}^{(s)}(p) = \not{p} + m,$$

Operate $u^{(s')}(p)$ on both sides, and use $\not{p}u^{(s')}(p) = mu^{(s')}(p)$ and $\bar{u}^{(s)}(p)u^{s'}(p) = 2m\delta_{s,s'}$ to get the identity $2mu^{(s)}(p) = 2mu^{(s)}(p)$. For the positrons change the signs.

Problem 2.24. Show that the matrix $S = \cosh(\chi/2) + \sinh(\chi/2)\gamma^0\gamma^3$, where the rapidity χ is defined in Problem 2.6: $\tanh \chi = v$. You will need some half angle identities for hyperbolic functions: $\sinh x = 2\sinh(x/2)\cosh(x/2)$, just like trigonometry, and

$\cosh x = \cosh^2(x/2) + \sinh^2(x/2)$, while $\cosh^2 x - \sinh^2 x = 1$. It was noted above that three-dimensional rotations for spin 1/2 vectors involve half angles. It is interesting that the Lorentz transformations for Dirac spinors also involve half angles — only the angle is rapidity, and the functions are hyperbolic.

Solution 2.24. Lorentz transformation of the Dirac equation. Equation (2.178) reads:

$$S = \sqrt{(\gamma+1)/2} + \sqrt{(\gamma-1)/2}\gamma^0\gamma^3$$
$$= \sqrt{(\cosh\chi+1)/2} + \sqrt{(\cosh\chi-1)/2}\gamma^0\gamma^3;$$

then use the identities $\cosh\chi + 1 = \cosh^2(\chi/2) + \sinh^2(\chi/2) + 1 = 2\cosh^2(\chi/2)$, and $\cosh\chi - 1 = \cosh^2(\chi/2) + \sinh^2(\chi/2) - 1 = 2\sinh^2(\chi/2)$ to obtain the desired result.

Problem 2.25. Use the matrix for S given by Equation (2.180) to 'Lorentz boost' the rest frame spinor:

$$u^{(1)}(m) = \sqrt{2m} \begin{pmatrix} 1 \\ 0 \\ 0 \\ 0 \end{pmatrix}. \tag{2.235}$$

The final spinor has momentum $p_z = v\gamma m$ Even the normalization comes out correctly!

Solution 2.25. Lorentz boost of a Dirac spinor at rest. Use the matrix given in Equation (2.180). Then

$$Su^{(1)}(m) = \sqrt{\frac{\gamma+1}{2}}\sqrt{2m} \begin{pmatrix} 1 \\ 0 \\ \sqrt{\frac{\gamma-1}{\gamma+1}} \\ 0 \end{pmatrix} = \sqrt{E+m} \begin{pmatrix} 1 \\ 0 \\ \frac{p}{E+m} \\ 0 \end{pmatrix}.$$

Problem 2.26. Show that the time reversal operation Equation (2.181) operating on the positive energy spinors give the positron spinors, namely,

$$S_T u^{(1)}(\vec{p}) = iv^{(2)}(-\vec{p}), \tag{2.236}$$

and

$$S_T u^{(2)}(\vec{p}) = iv^{(1)}(-\vec{p}).\qquad(2.237)$$

Solution 2.26. Equations (2.236) and (2.237) should be:

$$S_T u^{(1)}(\vec{p}) = iv^{(2)}(-\vec{p});\qquad(1)$$

and

$$S_T u^{(2)}(\vec{p}) = iv^{(1)}(-\vec{p}).\qquad(2)$$

Problem 2.27. Prove the Gordon decomposition of the vector current, Equation (2.202), using the procedure outlined in the text.

Solution 2.27. Gordon decomposition proof goes as outlined in the text. Work from the right-hand side, use the anti-commutation rules and the Dirac equation.

$$\frac{e}{2m}(\bar{u}(p_f)(p_f + p_i)^\mu u(p_i) + \bar{u}(p_f)i\sigma^{\mu\nu}(p_f - p_i)_\nu u(p_i))$$

$$= \frac{e}{2m}(\bar{u}(p_f)(p_f + p_i)^\mu u(p_i) - \bar{u}(p_f)$$

$$\times (-\not{p}_f\gamma^\mu + p_f^\mu - \gamma^\mu\not{p}_i + p_i^\mu)u(p_i)$$

$$= e\bar{u}(p_f)\gamma^\mu u(p_i).$$

Problem 2.28. The charge conjugate spinor is $\psi_C = i\gamma^2\psi^*$. Show that $\bar{\psi}_C\psi_C = -\bar{\psi}\psi$. Show that the TCP product flips the electron spin: $iu^{(2)*}(p) = i\gamma^0\gamma^3\gamma^1 \times u^{(1)}(p)$, and that the same operation works for the positron spinors.

Solution 2.28.

$$\psi_C = i\gamma^2\psi^*; \quad \psi_C^\dagger = -i\tilde{\psi}\gamma^{2\dagger};$$

where $\tilde{\psi}$ is a row vector that is the transpose of ψ without complex conjugation. Then

$$\bar{\psi}_C\psi_C = \tilde{\psi}\gamma^0\gamma^2\gamma^2\psi^* = -\tilde{\psi}\gamma^0\psi^* = -\bar{\psi}\psi;$$

because $\bar{\psi}\psi$ is real. Equation (2.209) gives:

$$PTC = \gamma^5\gamma^2 = \begin{pmatrix} 0 & i & 0 & 0 \\ -i & 0 & 0 & 0 \\ 0 & 0 & 0 & -i \\ 0 & 0 & i & 0 \end{pmatrix};$$

$$\gamma^5\gamma^2 u^{(1)}(p) = \sqrt{E+m}\begin{pmatrix} 0 \\ -i \\ \dfrac{-ip_x + p_y}{E+m} \\ \dfrac{ip_z}{E+m} \end{pmatrix} = -iu^{(2)*}(p).$$

Chapter 3

Problem 3.1. Use the commutation rules and the hypothesis $a\,|0\rangle = 0$ to obtain the matrix elements $a\,|n\rangle = \sqrt{n}\,|n-1\rangle$, $a^\dagger\,|n\rangle = \sqrt{n+1}\,|n+1\rangle$, and the number operator $N = a^\dagger a$, $N\,|n\rangle = n\,|n\rangle$. Show that $|n\rangle = \frac{(a^\dagger)^n}{\sqrt{n!}}\,|0\rangle$.

Solution 3.1. To calculate the matrix elements, use $\left[a, a^\dagger\right] = 1$, $a\,|0\rangle = 0$, $H = \hbar\omega(a^\dagger a + 1/2)$, and $H\,|n\rangle = E_n\,|n\rangle$.

$$Ha^\dagger\,|n\rangle = (E_n + \hbar\omega)a^\dagger\,|n\rangle\,;\quad Ha\,|n\rangle = (E_n - \hbar\omega)a\,|n\rangle.$$

Hence

$$a^\dagger\,|n\rangle = \lambda_{n+1}\,|n+1\rangle\,;\quad a\,|n\rangle = \lambda_n^*\,|n-1\rangle\,;$$

$$\langle n|\,a = \lambda_{n+1}^*\,\langle n+1|\,;\quad \langle n|\,a^\dagger = \lambda_n\,\langle n-1|\,.$$

Then with the kets normalized to one, the commutator $\left[a, a^\dagger\right] = 1$ gives a difference equation:

$$\langle n|\,aa^\dagger\,|n\rangle - \langle n|\,a^\dagger a\,|n\rangle = |\lambda_{n+1}|^2 - |\lambda_n|^2 = 1\,;$$

$$\lambda_n = \sqrt{n}\,;\quad a\,|0\rangle = 0\,;\quad a^\dagger a\,|n\rangle = n\,|n\rangle\,.$$

$$a^\dagger\,|0\rangle = |1\rangle\,;\quad (a^\dagger)^2\,|0\rangle = \sqrt{2}\,|2\rangle\,;\quad (a^\dagger)^3\,|0\rangle = \sqrt{3\times 2}\,|3\rangle\,;\ \textit{etc.}$$

Problem 3.2. Use $a\,|0\rangle = 0$ in operator form to obtain a first-order differential equation for the ground state wave function $\psi_0(x)$. Solve for $\psi_0(x)$, and normalize it. The Gaussian normalization integral is $\int dx e^{-\frac{x^2}{2\sigma^2}} = \sqrt{2\pi}\sigma$.

Solution 3.2. In operator form $a = x/x_0 + x_0 d/dx$, so $a\psi_0(x) = 0$ gives:

$$\frac{d\psi_0}{dx} = -\frac{x}{x_0^2}\psi_0; \quad \psi_0 = Ce^{-x^2/(2x_0^2)}; \quad \int dx|\psi_0|^2 = 1 \rightarrow C^2 = \frac{1}{\sqrt{\pi}x_0}.$$

Problem 3.3. For the harmonic oscillator $\int dx\psi_n^*(x)x\psi_n(x) = \int dx\psi_n^*(x)\,(d/dx)\psi_n(x) = 0$ for all energy eigenfunctions. The eigenfunctions have definite parity, and the operators have odd parity. Calculate $\int dx\psi_0^*(x)x^2\psi_0(x)$ and $-\int dx\psi_0^*(x)\frac{d^2}{dx^2}\psi_0(x)$, and show that the uncertainty product $\Delta x\Delta p = \hbar/2$, the minimum uncertainty. Kinetic and potential energy on the average share equally in the energy of the state $E_n = \hbar\omega(n + 1/2)$. Use this to obtain a formula for any energy eigenvalue $\langle \Delta x\Delta p\rangle_n = \hbar(n + 1/2)$. The general formula with $n = 0$ serves as a check on the wave function calculations. Harmonic oscillator energy eigenfunctions do not oscillate. They are constant in time.

Solution 3.3.

$$\langle x^2\rangle = \frac{1}{\sqrt{\pi}}\int dx x^2 e^{-x^2/x_0^2} = \frac{x_0^2}{2};$$

$$\langle p^2\rangle = -\frac{\hbar^2}{\sqrt{\pi}x_0^3}\int dx e^{-x^2/x_0^2}\left(\frac{x^2}{x_0^2} - 1\right) = \frac{\hbar^2}{2x_0^2}.$$

$$(\Delta p)^2(\Delta x)^2 = \frac{\hbar^2}{4}; \quad \Delta p\Delta x = \frac{\hbar}{2}.$$

$$\langle p^2\rangle/(2m) = \hbar\omega(n + 1/2)/2 = k/2\langle x^2\rangle;$$

$$\langle p^2\rangle\langle x^2\rangle = \hbar^2(n + 1/2)^2.$$

Problem 3.4. It is possible to construct an oscillating Gaussian packet that moves back and forth with the classical frequency, and maintains a constant minimum uncertainty. Such a packet contains a distribution of energies, and is called a *coherent state*. Coherent states have a number of unusual properties, and are an important part of quantum optics.

(a) Take a minimum packet, the ground state wave function, and displace it from the origin by a distance $x \to (x - a)$. This is the wave function at $t = 0$:

$$\psi(x, t = 0) = \left(\frac{1}{\sqrt{\pi}x_0}\right)^{1/2} e^{\frac{(x-a)^2}{2x_0^2}} = \sum_{n=0}^{\infty} A_n \psi_n(x), \qquad (3.244)$$

The expansion coefficients A_n are constants, the eigenfunctions of the Hamiltonian are $\psi_n(x)$, and the time dependent wave function $\psi(x, t)$ is obtained by multiplying each eigenfunction by $e^{-i\omega(n+1/2)t}$. The coefficients A_n may be determined using the orthonormality of the eigenfunctions:

$$A_n = \int_{-\infty}^{\infty} dx \psi_n^*(x)\psi(x, 0)$$

$$= \frac{1}{\sqrt{\pi}} \left(\frac{1}{2^n n!}\right)^{1/2} \int d\xi H_n(\xi) e^{(-\xi^2 + \xi\xi_0 - \xi_0^2/2)}, \qquad (3.245)$$

where the dimensionless variable $\xi = x/x_0$, and the constant $\xi_0 = a/x_0$. The integral looks hopeless, but can be evaluated by using the generating function, and looking at a the following expression:

$$\int d\xi e^{(-s^2 + 2s\xi - \xi^2 + \xi\xi_0 - \xi_0^2/2)} = \sum_{n=0}^{\infty} \int d\xi \frac{H_n(\xi)s^n}{n!} e^{(-\xi^2 + \xi\xi_0 - \xi_0^2/2)}.$$

$$(3.246)$$

The author is indebted to Schiff's '*Quantum Mechanics*' for this analysis. The integral on the left can be evaluated by completing the square in the exponential, and using the Gaussian integral formula. Show that the result is

$$\int d\xi e^{(-s^2 + 2s\xi - \xi^2 + \xi\xi_0 - \xi_0^2/2)} = \sqrt{\pi} e^{-\xi_0^2/4} e^{s\xi_0}. \qquad (3.247)$$

Then expand $e^{s\xi_0}$ in a Taylor series and match orders of s on both sides of the equation to obtain:

$$A_n = \left(\frac{1}{2^n n!}\right)^{1/2} e^{-\xi_0^2/4} (\xi_0)^n. \qquad (3.248)$$

This gets substituted into Equation (3.244). Add in the energy-time phases to get $\psi(x, t)$. Now use the generating function in reverse to eliminate $H_n(\xi)$, by identifying $s = \frac{\xi_0 e^{-i\omega t}}{2}$. Show that the oscillating wave function is now a pure exponential:

$$\psi(x,t) = \left(\frac{x_0}{\sqrt{\pi}}\right)^{1/2} \exp\left(-\xi^2/2 - \frac{\xi_0^2}{4}(1 + \cos(2\omega t)) + \xi\xi_0 \cos(\omega t)\right.$$

$$\left. + i\left(\frac{\xi_0^2 \sin(2\omega t)}{4} - \xi_0\xi \sin(\omega t) - \omega t/2\right)\right). \qquad (3.249)$$

Show that $\psi(x, t = 0)$ is an eigenfunction of the lowering (destruction) operator $a = (Q + iP)/(\sqrt{2})$ with eigenvalue $\xi_0/\sqrt{2}$. Then obtain the magnitude squared of the oscillating wave function:

$$|\psi(x,t)|^2 = \frac{1}{x_0\sqrt{\pi}}e^{-\frac{(x-a\cos(\omega t))^2}{x_0^2}}. \qquad (3.250)$$

This is a Gaussian shaped minimum uncertainty wave packet that oscillates back and forth about the origin with amplitude a. It was obtained simply by displacing the ground state wave function a distance a from the origin, and letting it loose.

Problem 3.4. (b) Coherent states in ket language. Define the eigenket $|\lambda\rangle$:

$$|\lambda\rangle = e^{-|\lambda|^2/2}e^{\lambda a^\dagger}|0\rangle, \qquad (3.251)$$

where the eigenvalue λ is in general a complex number, since the operator a is not Hermitian. Show that $|\lambda\rangle$ is an eigenket of a:

$$a|\lambda\rangle = \lambda|\lambda\rangle. \qquad (3.252)$$

Use the power series expansion of the exponential to show that:

$$|\lambda\rangle = e^{-|\lambda|^2/2}\sum_{n=0}^{\infty}\frac{\lambda^n |n\rangle}{\sqrt{n!}}. \qquad (3.253)$$

By comparing this power series with Equation (3.244), we have $\xi_0 = \lambda\sqrt{2}$. Defining $\langle\lambda|$ from the power series for the ket:

$$\langle\lambda| = e^{-|\lambda|^2/2}\sum_{n'=0}^{\infty}\frac{(\lambda^*)^{n'} \langle n'|}{\sqrt{n'!}}. \qquad (3.254)$$

Use these forms to show that $|\lambda\rangle$ is normalized:

$$\langle\lambda|\lambda\rangle = 1. \tag{3.255}$$

The normalization can also be checked from the conjugate form of Equation (3.251), namely,

$$\langle\lambda| = e^{-|\lambda|^2/2} \langle 0| e^{\lambda^* a}. \tag{3.256}$$

The proof of this assertion is deferred until Wick's theorem has been introduced. See Problem 3.10.

Derive the average energy level quantum number $\bar{n} = |\lambda|^2$ using Equations (3.253) and (3.254), and the operator $N = a^\dagger a$. Show that the eigenvalues are Poisson distributed:

$$Pr\{n\} = \frac{\bar{n}^n e^{-\bar{n}}}{n!}. \tag{3.257}$$

Solution 3.4. (a) Complete the square on the exponential:

$$- s^2 + 2s\xi - \xi^2 + \xi\xi_0 - \xi_0^2/2$$

$$= -(\xi^2 - 2(s + \xi_0/2)\xi + (s + \xi_0/2)^2 + s^2 + \xi_0^2/2 - (s + \xi_0/2)^2)$$

$$= -(\xi - (s + \xi_0/2))^2 + s\xi_0 - \xi_0^2/4; \int d\xi e^{-(\xi-(s+\xi_0/2))^2} = \sqrt{\pi}.$$

This gives Equation (3.247); specifically:

$$\int d\xi e^{-s^2+2s\xi-\xi^2+\xi\xi_0-\xi_0^2/2}$$

$$= \sum_{n=0}^{\infty} \int d\xi H_n(\xi)\frac{s^n}{n!} e^{-\xi^2+\xi\xi_0-\xi_0^2/2}$$

$$= \sqrt{\pi}e^{-\xi_0^2/4} \times e^{s\xi_0}.$$

Then expanding:

$$e^{s\xi_0} = \sum_{n=0}^{\infty} \frac{(\xi_0)^n}{n!}s^n;$$

and matching powers of s gives Equation (3.248):

$$A_n = \left(\frac{1}{2^n n!}\right)^{1/2} e^{-\xi_0^2/4}\xi_0^n.$$

Then the time-dependent wave function becomes:

$$\psi(x,t) = \frac{e^{-\xi_0^2/4}}{\pi^{1/4}} e^{-i\omega t/2} \sum_{n=0}^{\infty} \frac{\xi_0^n}{2^n n!} H_n(\xi) e^{-\xi^2/2} e^{-in\omega t}.$$

Now identify in the generating function formula $s = \xi_0 e^{-i\omega t}/2$. This works because the energy is proportional to n — it is a harmonic oscillator. With this choice, the wave function is entirely an exponential, and the imaginary part cancels when calculating $\psi^*\psi$:

$$\psi(x,t) = \frac{1}{\sqrt{x_0}\pi^{1/4}} e^{-\xi_0^2(1+e^{-2i\omega t})/4 - i\omega t/2 - \xi^2/2 + \xi\xi_0 e^{-i\omega t}};$$

$$\psi^*\psi = \frac{1}{x_0\sqrt{\pi}} e^{-(\xi - \xi_0 \cos(\omega t))^2}.$$

$\psi^*\psi$ oscillates back and forth in time with the classical frequency ω and the amplitude $a = \xi_0 x_0$, the initial displacement of the ground state wave function. The packet does not spread in time. The wave function at $t = 0$ is an eigenfunction of the lowering operator a:

$$a = \frac{Q + iP}{\sqrt{2}} = \frac{x}{x_0} + \frac{x_0 d}{dx};$$

$$a\psi(x,t=0) = \left(\frac{1}{\sqrt{\pi}x_0}\right)^{1/2} \left(\frac{x}{x_0} - \frac{x-a}{x_0}\right) e^{-(x-a)^2/(2x_0^2)} = \frac{a}{x_0}\psi.$$

Hopefully the use of a for both the lowering operator and the initial displacement does not cause confusion. This remarkable analysis was first done by Schrödinger soon after he solved the harmonic oscillator problem.

Solution 3.4. (b) Do 3.253 first.

$$|n\rangle = \frac{(a^\dagger)^n |0\rangle}{\sqrt{n!}}; \quad so \quad |\lambda\rangle = e^{-|\lambda|^2/2} \sum_{n=0}^{\infty} \frac{\lambda^n |n\rangle}{\sqrt{n!}}.$$

Then

$$a|\lambda\rangle = e^{-|\lambda|^2/2} \sum_{n=1}^{\infty} \frac{\lambda^n |n-1\rangle}{\sqrt{(n-1)!}} = e^{-|\lambda|^2/2} \sum_{n=0}^{\infty} \frac{\lambda^{n+1} |n\rangle}{\sqrt{n!}} = \lambda|\lambda\rangle.$$

The normalization is obvious from Equations (3.253) and (3.254), given the orthonormality of the kets: $\langle n'|n\rangle = \delta_{n',n}$. The operator $N = aa^\dagger$ is diagonal: $\langle n'|N|n\rangle = n\delta_{n,n'}$. So

$$\langle\lambda|N|\lambda\rangle = e^{-|\lambda|^2}\sum_{n=0}^{\infty}\frac{n|\lambda|^{2n}}{\sqrt{n!}} = e^{-|\lambda|^2}\sum_{n=1}^{\infty}\frac{|\lambda|^{2n}}{\sqrt{(n-1)!}} = |\lambda|^2.$$

For the Poisson distribution, calculate $|\langle n|\lambda\rangle|^2$:

$$\langle n|\lambda\rangle = e^{-|\lambda|^2/2}\frac{\lambda^n}{\sqrt{n!}}; \quad |\langle n|\lambda\rangle|^2 = e^{-|\lambda|^2}\frac{(|\lambda|^2)^n}{n!}.$$

$|\lambda|^2 = \bar{n}$.

Problem 3.5. If someone told you that the complete angular momentum algebra can be derived from two independent, non-interacting one-dimensional harmonic oscillators, you might be surprised, but you might also guess that it is not too difficult to prove. If you know that something can be done, it is often easy to do it. Schwinger was the first to show this. Take the angular momentum matrix elements in the following forms:

$$J_+ = J_x + iJ_y; J_+|j,m\rangle$$
$$= \sqrt{(j-m)(j+m+1)}\,|j,m+1\rangle, \tag{3.258}$$
$$J_- = J_+^\dagger = J_x - iJ_y; J_-|j,m\rangle$$
$$= \sqrt{(j+m)(j-m+1)}\,|j,m-1\rangle, \tag{3.259}$$
$$J_z|j,m\rangle = m|j,m\rangle, \tag{3.260}$$
$$J^2 = (J_+J_- + J_-J_+)/2 + J_z^2; J^2|j,m\rangle = j(j+1)|j,m\rangle, \tag{3.261}$$
$$[J_+, J_-] = 2J_z, \tag{3.262}$$

where $-j \leq m \leq +j, 2j+1$ values, j integer or half integer, including 0.

Call the two independent harmonic oscillators a and b, each perfectly ordinary, so $[a, a^\dagger] = 1$, and $[b, b^\dagger] = 1$, and everything else commutes, in particular $[a, b^\dagger] = [a, b] = 0$. Quantum numbers are

n_a and n_b, and $a\,|n_a\rangle = \sqrt{n_a}\,|n_a - 1\rangle$, etc. Try $J_+ = a^\dagger b$. Show that the matrix elements for J_+ and J_- are correct if $n_a = j + m$ and $n_b = j - m$. Calculate the matrix elements of J_z and J^2. With $j = (n_a + n_b)/2$ and $m = (n_a - n_b)/2$, show that all representations and eigenvalues for J^2 and J_z are accounted for. Make a table of n_a, n_b, j, m for $j = 0, 1/2, 1, 3/2$. The matrices in the n_a, n_b representation are awkward. To begin with, the labels are upside down, since $n = 0$ is at the top of the infinite matrix, and the raising operator goes down one notch. Nevertheless, one can use the matrices. Write down $2J_z = a^\dagger a - b^\dagger b$, and find the eigenvectors and eigenvalues, for spin $1/2$. *Hint*:

$$\left[\begin{pmatrix} 0 & 0 \\ 0 & 1 \end{pmatrix} - \begin{pmatrix} 0 & 0 \\ 0 & 1 \end{pmatrix} \right] \begin{pmatrix} 0 \\ 1 \end{pmatrix} \begin{pmatrix} 1 \\ 0 \end{pmatrix} \tag{3.263}$$

$a^\dagger a$ is the first 2×2 matrix, and $n_a = 1$ is the ket on the left; $n_b = 0$ is on the right. The a matrices do not see the b matrices or the b kets.

Solution 3.5.

$$J_+\,|n_a, n_b\rangle = ab^\dagger\,|n_a, n_b\rangle = \sqrt{(n_a + 1)n_b}\,|n_a + 1, n_b - 1\rangle\,;$$
$$J_-\,|n_a, n_b\rangle = \sqrt{n_a(n_b + 1)}\,|n_a - 1, n_b + 1\rangle\,;$$

match the angular momentum matrix elements if $n_a = j + m, n_b = j - m$. For J_z:

$$J_+ J_-\,|n_a, n_b\rangle = n_a(n_b + 1)\,|n_a, n_b\rangle\,;$$
$$J_- J_+\,|n_a, n_b\rangle = n_b(n_a + 1)\,|n_a, n_b\rangle\,;$$
$$2J_z = n_a - n_b; \quad J_z = (n_a - n_b)/2.$$
$$J^2 = (J_+ J_- + J_- J_+)/2 + J_z^2 = \frac{n_a + n_b}{2} + \frac{(n_a + n_b)^2}{4} = j(j + 1);$$
$$j = (n_a + n_b)/2.$$

Table of Angular Momentum Quantum Numbers in terms of n_a and n_b harmonic oscillators

n_a	n_b	j	m_j
0	0	0	0
1	0	1/2	+1/2
0	1	1/2	−1/2
1	1	1	0
2	0	1	+1
0	2	1	−1
2	1	3/2	+1/2
1	2	3/2	−1/2
3	0	3/2	+3/2
0	3	3/2	−3/2

Problem 3.6. Using the definition of the field amplitude $\Phi(x)$ in Equation (3.43), show that the commutation rules Equations (3.45)–(3.47) give the desired result $[\Phi(x), \dot{\Phi}(y)] = i\delta(x - y)$.

Solution 3.6. Dropping the terms where the operators commute, we have:

$$\left[\Phi(\vec{x}, t), \dot{\Phi}(\vec{y}, t)\right] = \frac{i}{(2\pi)^6} \int \frac{d^3k}{2\omega} \int \frac{d^3k'}{2}$$
$$\times \left\{ \left[a(\vec{k}), a^\dagger(\vec{k}')\right] e^{i(\vec{k}\cdot\vec{x} - \vec{k}'\cdot\vec{y})} e^{-i(\omega-\omega')t} \right.$$
$$\left. - \left[a^\dagger(\vec{k}), a(\vec{k}')\right] e^{i(\vec{k}'\cdot\vec{y} - \vec{k}\cdot\vec{x})} e^{-i(\omega'-\omega)t} \right\}.$$

Flip the sign and change the order of the second commutator and use the rule $\left[a(\vec{k}), a^\dagger(\vec{k}')\right] = (2\pi)^3 2\omega\delta(\vec{k} - \vec{k}')$ to obtain:

$$\left[\Phi(\vec{x}, t), \dot{\Phi}(\vec{y}, t)\right] = \frac{i}{(2\pi)^3} \int \frac{d^3k}{2} \left\{ e^{i\vec{k}\cdot(\vec{x}-\vec{y})} + e^{-i\vec{k}\cdot(\vec{x}-\vec{y})} \right\}.$$

This gives the desired result. The δ function is even, so the two terms add, cancelling the 1/2.

Problem 3.7. Obtain the formula for the Hamiltonian Equation (3.54).

Solution 3.7. Assemble Equations (3.51), (3.52), and (3.53). The time-dependent terms are multiplied by

$$-\frac{1}{4} + \frac{\vec{k}^2}{4\omega^2} + \frac{m^2}{4\omega^2} = 0.$$

The time-independent terms are multiplied by

$$\frac{1}{4} + \frac{\vec{k}^2}{4\omega^2} + \frac{m^2}{4\omega^2} = \frac{1}{2}.$$

Problem 3.8. Charged field. Part 1, Klein–Gordon equation. There are four classical fields in this problem, all with the same mass, and all satisfying the Klein–Gordon equation. Show that the Lagrange density for two non-interacting real fields $\phi_1(x)$ and $\phi_2(x)$:

$$\mathcal{L} = (\partial_\mu \phi_1 \partial^\mu \phi_1 - m^2 \phi_1^2)/2 + (\partial_\mu \phi_2 \partial^\mu \phi_2 - m^2 \phi_2^2)/2, \quad (3.264)$$

becomes one equation in ϕ and ϕ^*:

$$\mathcal{L} = \partial_\mu \phi^* \partial^\mu \phi - m^2 \phi^* \phi, \quad (3.265)$$

where $\phi = (\phi_1 + i\phi_2)/\sqrt{2}$ and $\phi^* = (\phi_1 - i\phi_2)/\sqrt{2}$. Treat ϕ and ϕ^* as independent variables in Lagrange's equations. Show that the four-dimensional current j^μ is conserved:

$$\partial_\mu(\phi^* \partial^\mu \phi - (\partial^\mu \phi^*)\phi) = 0. \quad (3.266)$$

Define the charge density ρ as the elementary charge times the time component of j^μ:

$$\rho = -iej^0(x) = -ie(\phi^* \partial\phi/\partial t - (\partial\phi^* \partial t)\phi). \quad (3.267)$$

As mentioned in the text, the quantized field (no longer Hermitian) has to couple operators for Φ and $\Phi^\dagger$, thus,

$$\Phi(x) = \frac{1}{(2\pi)^3} \int \frac{d^3k}{2\omega} (a(\vec{k})e^{-ik\cdot x} + b^\dagger(\vec{k})e^{ik\cdot x}). \quad (3.268)$$

The a's and b's satisfy the usual commutation relations Equations (3.45)–(3.47), while a and $a^\dagger$ commute with b and $b^\dagger$. Show that after normal ordering the total charge Q is given by:

$$Q = \int d^3x \rho(x) = \frac{e}{(2\pi)^3} \int \frac{d^3k}{2\omega} (a^\dagger(\vec{k})a(\vec{k}) - b^\dagger(\vec{k})b(\vec{k})). \quad (3.269)$$

Normal ordering reversed the order of the b term, but the minus sign came from the definition of ρ. The time dependent terms in the expansion also vanish because ρ is the difference between two terms.

Solution 3.8. Classical fields commute.

$$\phi_1 = \frac{\phi + \phi^*}{\sqrt{2}}; \quad \phi_2 = \frac{\phi - \phi^*}{i\sqrt{2}};$$

$$\phi_1^2 + \phi_2^2 = 2\phi^*\phi; \quad \partial_\mu\phi_1\partial^\mu\phi_1 + \partial_\mu\phi_2\partial^\mu\phi_2 = 2\partial_\mu\phi^*\partial^\mu\phi.$$

Hence

$$\mathcal{L} = \partial_\mu\phi^*\partial^\mu\phi - m^2\phi^*\phi; \quad \frac{\partial}{\partial x^\nu}\frac{\partial\mathcal{L}}{\partial(\partial_\nu\phi)} = \partial_\nu\partial^\nu\phi^* = -m^2\phi^*.$$

Same equation for ϕ. So multiply one by ϕ, and the other by ϕ^*, and subtract:

$$\phi^*\partial_\nu\partial^\nu\phi - (\partial_\nu\partial^\nu\phi^*)\phi = 0.$$

This is the same as the current conservation equation:

$$\partial_\nu(\phi^*\partial^\nu\phi - (\partial^\nu\phi^*)\phi) = 0.$$

The extra terms cancel. Write down the four Fourier expansions of the fields:

$$\Phi(x) = \frac{1}{(2\pi)^3}\int\frac{d^3k}{2\omega}\left(a(\vec{k})e^{i(\vec{k}\cdot\vec{x}-\omega t)} + b^\dagger(\vec{k})e^{-i(\vec{k}\cdot\vec{x}-\omega t)}\right);$$

$$\Phi^\dagger(x) = \frac{1}{(2\pi)^3}\int\frac{d^3k}{2\omega}\left(a^\dagger(\vec{k})e^{-i(\vec{k}\cdot\vec{x}-\omega t)} + b(\vec{k})e^{i(\vec{k}\cdot\vec{x}-\omega t)}\right);$$

$$\dot\Phi(x) = \frac{i}{(2\pi)^3}\int\frac{d^3k}{2}\left(-a(\vec{k})e^{i(\vec{k}\cdot\vec{x}-\omega t)} + b^\dagger(\vec{k})e^{-i(\vec{k}\cdot\vec{x}-\omega t)}\right);$$

$$\dot\Phi^\dagger(x) = \frac{i}{(2\pi)^3}\int\frac{d^3k}{2}\left(a^\dagger(\vec{k})e^{-i(\vec{k}\cdot\vec{x}-\omega t)} - b(\vec{k})e^{i(\vec{k}\cdot\vec{x}-\omega t)}\right);$$

The operator structure of the charge density is:

$$\Phi^\dagger\dot\Phi - \dot\Phi^\dagger\Phi \sim -a^\dagger a - ba + a^\dagger b^\dagger + bb^\dagger - a^\dagger a - a^\dagger b^\dagger + ba + bb^\dagger.$$

The ba and $a^\dagger b^\dagger$ terms cancel. The $a^\dagger a$ term is normal ordered, but the $bb^\dagger$ term has to be reversed. This does not change the sign, since this is a boson field. So the total charge operator is:

$$Q = -ie \int d^3x (\Phi^\dagger \dot\Phi - \dot\Phi^\dagger \Phi) = \frac{e}{(2\pi)^6} \int d^3x \int \frac{d^3k'}{2\omega'} \int \frac{d^3k}{2}$$
$$\times \left\{ -2a^\dagger(\vec{k}')a(\vec{k}) e^{i(\vec{k}'-\vec{k})\cdot\vec{x}} e^{-i(\omega'-\omega)t} \right.$$
$$\left. + 2b^\dagger(\vec{k})b(\vec{k}') e^{i(\vec{k}'-\vec{k})\cdot\vec{x}} e^{-i(\omega'-\omega)t} \right\}. \tag{1}$$

The space integral gives a $\delta(\vec{k}' - \vec{k})$, then $\int d^3k$ removes the time dependence, and gives the desired result. The a field and the b field have opposite charges, and the net charge is the occupation number difference.

Problem 3.9. Charged field Part 2, Dirac equation. Repeat the calculation of the total charge for the Dirac operator $j^\mu = e\bar\Psi\gamma^\mu\Psi$. The result is the same, apart from a sum over spins. Note the differences in the calculation. $\Psi^\dagger\Psi$ has no time derivatives, and is not the difference of two quantities. The cancellation is taken care of by the orthogonality of the spinors, and the sign change on the $b^\dagger b$ term is achieved because b and $b^\dagger$ anticommute.

Solution 3.9. Evaluate $Q = \int d^3x\rho$ using the Dirac equation. For the Dirac field $\rho = e\psi^\dagger\psi$, and

$$\psi(x) = \frac{1}{2\pi^3} \int \frac{d^3p}{2E} \sum_{s=1,2} \left\{ a_s(\vec{p})u_s(p)e^{-ip\cdot x} + b_s^\dagger(\vec{p})v_s(p)e^{ip\cdot x} \right\};$$

and

$$\psi^\dagger(x) = \frac{1}{2\pi^3} \int \frac{d^3p'}{2E'} \sum_{s'=1,2} \left\{ a_{s'}^\dagger(\vec{p}')u_{s'}^\dagger(p')e^{ip'\cdot x} \right.$$
$$\left. + b_{s'}(\vec{p}')v_{s'}^\dagger(p')e^{-ip'\cdot x} \right\}.$$

The reduction of $Q = e\int d^3x\psi^\dagger\psi$ proceeds in the same way as the Hamiltonian operator in the text, Equation (3.222). The components are:

$$e \int d^3x \psi^\dagger \psi = e \frac{1}{(2\pi)^6} \int d^3x \int \frac{d^3p}{2E} \int \frac{d^3p'}{2E'} \sum_{s=1,2} \sum_{s'=1,2}$$

$$\times \left\{ a_{s'}^\dagger(\vec{p}\,')u_{s'}^\dagger(p')a_s(\vec{p})u_s(p)e^{i(p'-p)\cdot x} \right.$$

$$+ a_{s'}^\dagger(\vec{p}\,')u_{s'}^\dagger(p')b_s^\dagger(\vec{p})v_s(p)e^{i(p+p')\cdot x}$$

$$+ b_{s'}(\vec{p}\,')v_{s'}(p')a_s(\vec{p})u_s(p)e^{-i(p+p')\cdot x}$$

$$\left. + b_{s'}(\vec{p}\,')v_{s'}(p')b_s^\dagger(\vec{p})v_s(p)e^{i(p-p')\cdot x} \right\}.$$

Space integration over the first and fourth terms gives $(2\pi)^3\delta(\vec{p}-\vec{p}\,')$, while the second and third terms give $(2\pi)^3\delta(\vec{p}+\vec{p}\,')$. Spinor products for one and four then give $2E\delta_{s,s'}$, and for two and three zero (Equations (3.220) and (3.221)). After normal ordering on the positron term the result is:

$$Q = e\frac{1}{(2\pi)^3} \int \frac{d^3p}{2E} \sum_{s=1,2} (a_s^\dagger(\vec{p})a_s(\vec{p}) - b_s^\dagger(\vec{p})b_s(\vec{p})).$$

The time-dependent terms are eliminated by the orthogonality of the spinors, and the sign change for the positrons comes from anticommutation for normal ordering.

Problem 3.10. Wick's theorem and the harmonic oscillator. We return to the problem of the normalization of coherent states of the harmonic oscillator:

$$|\lambda\rangle = e^{-|\lambda|^2/2}e^{\lambda a^\dagger}|0\rangle, \tag{3.270}$$

and

$$\langle\lambda| = e^{-|\lambda|^2/2}\langle 0|e^{\lambda^* a}. \tag{3.271}$$

Multiplying them together gives:

$$\langle\lambda|\lambda\rangle = e^{-\lambda^2}\langle 0|e^{\lambda a}e^{\lambda a^\dagger}|0\rangle, \tag{3.272}$$

where we have taken λ to be real, which it is in Schrödinger's solution. It is the vacuum expectation value of $e^{\lambda a}e^{\lambda a^\dagger}$ that must equal e^{λ^2} for the normalization. This is true, but proving it is an interesting exercise in the use of Wick's theorem for the reduction of vacuum expectation values. First, because $[a, a^\dagger] = 1$,

$$e^{\lambda a} e^{\lambda a^\dagger} = e^{\lambda(a+a^\dagger) + \lambda^2[a,a^\dagger]/2}, \tag{3.273}$$

and half of our problem is taken care of, so we are left with the task of showing that:

$$e^{\lambda^2/2} = \langle 0| \, e^{\lambda(a+a^\dagger)} \, |0\rangle . \tag{3.274}$$

Expand the exponential into products of the operators:

$$\begin{aligned}
e^{\lambda(a+a^\dagger)} &= 1 + \lambda(a + a^\dagger) + \lambda^2(a + a^\dagger)^2/2! + \lambda^3(a + a^\dagger)^3/3! \\
&\quad + \lambda^4(a + a^\dagger)^4/4! + \lambda^5(a + a^\dagger)^5/5! \\
&\quad + \lambda^6(a + a^\dagger)^6/6! + \cdots .
\end{aligned} \tag{3.275}$$

The vacuum expectation values of all odd powers of the operators vanish, so we are left with $0, 2, 4, 6, \ldots$ that match the powers of λ^2 in $e^{\lambda^2/2}$. Make a table matching coefficients of λ^2 in Equation (3.275) through sixth order. To have a non-vanishing vacuum expectation value, there must be an equal number of a's and $a^\dagger$'s; an a must be on the left, and an $a^\dagger$ on the right. So, $(a + a^\dagger)^4$ has two non-vanishing arrangements, namely, $aa^\dagger aa^\dagger$, and $aaa^\dagger a^\dagger$, and there are three individual pairings:

$$\langle 0| \, aa^\dagger aa^\dagger \, |0\rangle = \overbrace{aa^\dagger}\overbrace{aa^\dagger} = 1, \tag{3.276}$$

$$\langle 0| \, aaa^\dagger a^\dagger \, |0\rangle = \overbrace{a\overbrace{aa^\dagger}a^\dagger} = 1, \tag{3.277}$$

$$\langle 0| \, aaa^\dagger a^\dagger \, |0\rangle = \overbrace{aaa^\dagger a^\dagger} = 1. \tag{3.278}$$

All contractions are either one or zero. For the fourth-order term they add up to 3. The sixth-order term is not as daunting as it may seem. Gather your courage. Multiplying everything out gives 64 terms, but all but six have zero vacuum expectation value. The six terms are of the form $\langle 0| \, axxyya^\dagger \, |0\rangle$ with two a's and two $a^\dagger$'s in the middle, because $4!/(2!)^2 = 6$ (interchanging the two a's or the two $a^\dagger$'s doesn't change anything). One of these terms has no contribution, but the other five do. Work it out. Show that

$\langle 0| \, axxyya^\dagger \, |0\rangle = 15$, and that is exactly what is needed to match the two power series. Checking the formula to sixth order will suffice to prove the identity.

Solution 3.10. Wick's theorem and the harmonic oscillator. We are asked to show that:

$$\langle 0| \, e^{\lambda(a+a^\dagger)} \, |0\rangle = e^{\lambda^2/2}$$

using Wick's theorem that the vacuum expectation value of a product of creation and destruction operators is the sum of all possible contractions. For the harmonic oscillator $[a, a^\dagger] = 1$, so $\langle 0| \, [a, a^\dagger] \, |0\rangle = \langle 0| \, aa^\dagger \, |0\rangle = 1$. Expand both sides in power series:

$$1 + \lambda^2/2 + \lambda^4/(4 \times 2!) + \lambda^6/(8 \times 3!) + \cdots$$
$$= \langle 0| \, 1 + (a + a^\dagger)\lambda + (a + a^\dagger)^2 \lambda^2/2!$$
$$+ (a + a^\dagger)^3 \lambda^3/3! + (a + a^\dagger)^4 \lambda^4/4!$$
$$+ (a + a^\dagger)^5 \lambda^5/5! + (a + a^\dagger)^6 \lambda^6/6! + \cdots \, |0\rangle \, .$$

The vacuum expectation value of an odd number of operators vanishes. There must be an a on the left and an $a^\dagger$ on the right, and an equal number of a's and $a^\dagger$'s for a non-zero result. Matching the coefficients on the right with those on the left gives:

$$\langle 0| \, (a + a^\dagger)^2 \, |0\rangle = 1; \quad \langle 0| \, (a + a^\dagger)^4 \, |0\rangle = 3; \quad \langle 0| \, (a + a^\dagger)^6 \, |0\rangle = 15.$$

We use Wick's theorem to demonstrate the correctness of these equations. The terms are: λ^2

$$\langle 0| \, aa + a^\dagger a + aa^\dagger + a^\dagger a^\dagger \, |0\rangle = \langle 0| \, aa^\dagger \, |0\rangle = 1.$$

λ^4

$$\langle 0| \, aaa^\dagger a^\dagger + aa^\dagger aa^\dagger \, |0\rangle = 2 + 1 = 3.$$

The first term has two non-zero contractions.
λ^6

$$\langle 0| \, aa^\dagger a^\dagger aaa^\dagger + aa^\dagger aa^\dagger aa^\dagger + aaaa^\dagger a^\dagger a^\dagger + aaa^\dagger a^\dagger aa^\dagger$$
$$+ aaa^\dagger aa^\dagger a^\dagger + aa^\dagger aaa^\dagger a^\dagger \, |0\rangle = 0 + 1 + 6 + 2 + 4 + 2 = 15.$$

The third term has six non-zero contraction combinations. Write down the term six times and draw the different contraction links to convince yourself. Use this same procedure for the other terms.

Problem 3.11. Show that adding the four divergence $\partial_\mu A_0 F^{0\mu}$ to the Hamiltonian density for the electromagnetic field Equation (3.107), gives the desired form $\mathcal{H} = (\vec{E}^2 + \vec{B}^2)/2$. Remember that for a free field $\vec{\nabla} \cdot \vec{E} = 0$.

Solution 3.11. Equation (3.107) reads:

$$\mathcal{H} = -\frac{\vec{E}}{c^2} \cdot \frac{\partial \vec{A}}{\partial t} - \frac{\vec{E}^2}{2c^2} + \frac{\vec{B}^2}{2};$$

The desired form is:

$$\mathcal{H} = \frac{\vec{E}^2 + \vec{B}^2}{2}.$$

So we need to add a term:

$$\frac{\vec{E}}{c^2} \cdot \frac{\partial \vec{A}}{\partial t} + \frac{\vec{E}^2}{c^2} = \frac{1}{c^2} \vec{\nabla} V \cdot \left(\vec{\nabla} V + \frac{\partial \vec{A}}{\partial t} \right);$$

where we have used $\vec{E} = -\vec{\nabla} V - \partial \vec{A}/\partial t$. A suggested four divergence to add to the Hamiltonian density is:

$$\partial_\mu (A_0 \partial^0 A^\mu - A_0 \partial^\mu A^0) = \frac{\partial}{c^2 \partial x^i} \left(V \frac{\partial A^i}{\partial t} + V \frac{\partial V}{\partial x^i} \right).$$

The $\mu = 0$ term vanishes, and the sign of the second term has flipped to use the "usual" derivative $\partial/\partial x^i$. This expression and the one above it are the same. The extra term is $\vec{\nabla} \cdot \vec{E} = 0$ for a free field.

Problem 3.12. Gupta and Bleuler's argument, adopted from Otto Nachtmann's text. We learned that the gauge condition $k^\mu a_\mu = 0$ did not make sense as an operator equation, but it is possible to require:

$$k^\mu a_\mu \, |\text{physical photon state}\rangle = 0. \qquad (3.279)$$

Choose the z axis for the photon direction $\vec{k} = \omega \hat{z}$, and the allowed polarization directions are ϵ_x and ϵ_y. Define linear combinations of

$a_0(\vec{k})$ and $a_3(\vec{k})$ thus,

$$\alpha_0(\vec{k}) = (a_0(\vec{k}) - a_3(\vec{k}))/\sqrt{2}, \qquad (3.280)$$

$$\alpha_3(\vec{k}) = (a_0(\vec{k}) + a_3(\vec{k}))/\sqrt{2}, \qquad (3.281)$$

$$\alpha_1(\vec{k}) = a_1(\vec{k}), \qquad (3.282)$$

$$\alpha_2(\vec{k}) = a_2(\vec{k}). \qquad (3.283)$$

α_0 is chosen so that the gauge condition is:

$$\alpha_0(\vec{k}) \,|\text{physical photon state}\rangle = 0. \qquad (3.284)$$

$\alpha_3^\dagger(\vec{k})$ creates a rogue state that does not satisfy $\epsilon_\mu k^\mu = 0$. Using the commutation rules for $a_\mu(\vec{k})$ and $a_\mu^\dagger(\vec{k})$ Equations (3.113) and (3.114), work out the commutation rules for the α's and $\alpha^\dagger$'s. Show that a multiphoton state created by a string of $\alpha^\dagger$'s satisfies Equation (3.284) ($\alpha_0 \,|0\rangle = 0$), provided that there are no states created by an $\alpha_3^\dagger$.

Solution 3.12. $k^\mu a_\mu(\vec{k}) = 0$ is a gauge condition: $\partial_\mu A^\mu(x) = 0$ applied to the Fourier expansion of the vector potential. This operator requirement is not consistent with the commutation rules. This inconsistency, like the fact that the transverse polarization is not Lorentz invariant, has to be patched up, because it is not fundamentally fixable. The patch is to require $k^\mu a_\mu(\vec{k}) \,|\text{physical photon state}\rangle = 0$, and this can be done provided that one of four possible operations is forbidden. To isolate the forbidden operation, define linear combinations of the operators a_0 and a_3, with $k^\mu = (\omega, 0, 0, |\vec{k}|)$:

$$\alpha_0(\vec{k}) = (a_0(\vec{k}) - a_3(\vec{k}))/\sqrt{2}; \; k^\mu a_\mu = \omega(a_0 - a_3);$$

so our requirement $k^\mu a_\mu(\vec{k}) \,|\text{physical photon state}\rangle = 0$ is equivalent to $\alpha_0(\vec{k}) \,|\text{physical photon state}\rangle = 0$. The other linear combination $\alpha_3 = (a_0 + a_3)/\sqrt{2}$ destroys a rogue state, that must be forbidden. The commutation rules follow from the a's:

$$\left[a_\mu(\vec{k}), a_\nu^\dagger(\vec{k}') \right] = -(2\pi)^3 g_{\mu,\nu} 2\omega \delta(\vec{k} - \vec{k}').$$

Therefore:

$$\left[\alpha_0(\vec{k}), \alpha_0^\dagger(\vec{k}')\right] = \left[\alpha_3(\vec{k}), \alpha_3^\dagger(\vec{k}')\right] = 0;$$

$$\left[\alpha_0(\vec{k}), \alpha_3^\dagger(\vec{k}')\right] = -(2\pi)^3 2\omega\delta(\vec{k} - \vec{k}').$$

Operators for the two components of the polarization remain unchanged: $\alpha_1 = a_1$, $\alpha_2 = a_2$, $\left[\alpha_1, \alpha_1^\dagger\right] = \left[\alpha_2, \alpha_2^\dagger\right] = (2\pi)^3 2\omega\delta(\vec{k} - \vec{k}')$. $\alpha_0(\vec{k})$ commutes with all of the creation operators operating on the vacuum except for $\alpha_3^\dagger$, so if it is missing, then:

$$\alpha_0(\vec{k}) \left\{\alpha_1^\dagger(\vec{k}')\alpha_2^\dagger(\vec{k}'')\alpha_0^\dagger(\vec{k}''')\cdots\right\} |0\rangle = 0.$$

Problem 3.13. Beginning with $\vec{A}$ in Equation (3.128), go through the necessary steps to obtain Equation (3.134), the electromagnetic Hamiltonian:

$$H = \frac{1}{(2\pi)^3} \int \frac{d^3k}{2} \sum_s a^{\dagger(s)}(\vec{k})a^{(s)}(\vec{k}). \qquad (3.285)$$

Solution 3.13. This solution is outlined in the text. Equation (3.128) reads:

$$\vec{A}(\vec{x}, t) = \frac{1}{(2\pi)^3} \int \frac{d^3k}{2\omega} \sum_{s=1}^{2} \left\{\vec{\epsilon}^{(s)}a^{(s)}(\vec{k})e^{i(\vec{k}\cdot\vec{x}-\omega t)} \right.$$
$$\left. +\vec{\epsilon}^{*(s)}a^{\dagger(s)}(\vec{k})e^{-i(\vec{k}\cdot\vec{x}-\omega t)}\right\}.$$

The Hamiltonian density is:

$$\mathcal{H} = \frac{\vec{E}^2 + \vec{B}^2}{2}; \ \vec{E} = -\frac{\partial\vec{A}}{\partial t}; \ \vec{B} = \vec{\nabla} \times \vec{A}; \ H = \int d^3x\mathcal{H}.$$

To calculate $\mathcal{H}$ we need two expressions:

$$\frac{\partial\vec{A}}{\partial t} = \frac{i}{(2\pi)^3} \int \frac{d^3k}{2} \sum_{s=1}^{2} \left\{-\vec{\epsilon}^{(s)}a^{(s)}(\vec{k})e^{i(\vec{k}\cdot\vec{x}-\omega t)}\right.$$
$$\left. + \vec{\epsilon}^{*(s)}a^{\dagger(s)}(\vec{k})e^{-i(\vec{k}\cdot\vec{x}-\omega t)}\right\}.$$

$$\vec{\nabla} \times \vec{A} = \frac{i}{(2\pi)^3} \int \frac{d^3k}{2\omega} \sum_{s=1}^{2} \left\{ \vec{k} \times \vec{\epsilon}^{(s)} a^{(s)}(\vec{k}) e^{i(\vec{k}\cdot\vec{x}-\omega t)} \right.$$

$$\left. -\vec{k} \times \vec{\epsilon}^{*(s)} a^{\dagger(s)}(\vec{k}) e^{-i(\vec{k}\cdot\vec{x}-\omega t)} \right\}.$$

Let ϵ be real — linearly polarized light — and let $\vec{\epsilon}^{(1)} \times \vec{\epsilon}^{(2)} = \vec{k}/\omega$. Then $\vec{k} \times \vec{\epsilon}^{(1)} = \omega\vec{\epsilon}^{(2)}$, and $\vec{k} \times \vec{\epsilon}^{(2)} = -\omega\vec{\epsilon}^{(1)}$. The spin indices on ϵ and a are different, but otherwise $\vec{\nabla} \times \vec{A}$ and $\partial\vec{A}/\partial t$ look the same. We will leave the form above as-is, however, to explain the source of the cancellation. Square the electric field and integrate over space:

$$\int d^3x \vec{E}^2 = -\frac{1}{(2\pi)^6} \int d^3x \int \frac{d^3k}{2} \int \frac{d^3k'}{2} \sum_s \sum_{s'} \vec{\epsilon}^{(s)} \cdot \vec{\epsilon}^{(s')}$$

$$\times \left\{ a^{(s)}(\vec{k}) a^{(s')}(\vec{k'}) e^{i((\vec{k}+\vec{k'})\cdot\vec{x}-(\omega+\omega')t)} \right.$$

$$- a^{(s)}(\vec{k}) a^{\dagger(s')}(\vec{k'}) e^{i((\vec{k}-\vec{k'})\cdot\vec{x}-(\omega-\omega')t)}$$

$$- a^{\dagger(s)}(\vec{k}) a^{(s')}(\vec{k'}) e^{i((\vec{k}-\vec{k'})\cdot\vec{x}-(\omega-\omega')t)}$$

$$\left. + a^{\dagger(s)}(\vec{k}) a^{\dagger(s')}(\vec{k'}) e^{i((\vec{k}+\vec{k'})\cdot\vec{x}-(\omega+\omega')t))} \right\}.$$

$$\frac{1}{(2\pi)^3} \int d^3x e^{i(\vec{k}+\vec{k'})\cdot\vec{x}}$$

$$= \delta(\vec{k}+\vec{k'}); \quad \frac{1}{(2\pi)^3} \int d^3x e^{i(\vec{k}-\vec{k'})\cdot\vec{x}} = \delta(\vec{k}-\vec{k'}).$$

The free field Hamiltonian is time-independent, so the terms $e^{\pm 2i\omega t}$ must cancel, and they do because the sign change $\vec{k'} = -\vec{k}$ flips the sign of the $\vec{k'} \cdot \vec{\epsilon}$ terms in $\vec{B}^2$, leaving the two time independent parts in the middle to add to form the desired result Equation (3.133). Normal ordering gives (3.134).

Problem 3.14. Derive Equation (3.148).

Solution 3.14. Equation (3.148)

$$i\hbar \frac{\partial U_I(t,t_0)}{\partial t} = H_I(t) U_I(t,t_0); \quad \text{where } H_I(t) = e^{iH_0 t/\hbar} H_I e^{-iH_0(t)/\hbar};$$

and

$$U_I(t, t_0) = e^{iH_0 t/\hbar} U(t, t_0) e^{-iH_0 t_0/\hbar};$$

$$i\hbar \frac{\partial U(t, t_0)}{\partial t} = HU(t, t_0); \quad H = H_0 + H_I.$$

So:

$$i\hbar \frac{\partial U_I}{\partial t} = -H_0 U_I + e^{iH_0 t/\hbar} HU e^{-iH_0 t_0/\hbar}$$

$$= -e^{iH_0 t/\hbar} H_0 U e^{-iH_0 t/\hbar} + e^{iH_0 t/\hbar} H_0 U e^{-iH_0 t/\hbar}$$

$$+ e^{iH_0 t/\hbar} H_I U e^{-iH_0 t_0/\hbar} = H_I(t) U_I(t, t_0).$$

The first two terms cancel, and there are no steps that involve manipulating the exponentials of non-commuting operators.

Problem 3.15. Check the integration to obtain

$$\langle f| H_I |i\rangle = \frac{ZZ'e^2}{V} \frac{4\pi}{(K^2 + \alpha^2)}. \tag{3.286}$$

Solution 3.15. Start with Equation (3.158):

$$\langle f| H_I |i\rangle = \frac{ZZ'e^2}{V} \int d\Omega' \int_0^\infty r' dr' e^{(-\alpha + iK \cos\theta')r'}$$

$$= -\frac{2\pi ZZ'e^2}{V} \int_0^\infty r' dr' e^{-\alpha r'} \int_{-1}^1 d\cos\theta' e^{iKr' \cos\theta'};$$

$$= -\frac{2\pi ZZ'e^2}{V} \int_0^\infty dr' \left(\frac{e^{(-\alpha + iK)r'} - e^{(-\alpha - iK)r'}}{iK} \right)$$

$$= \frac{4\pi ZZ'e^2}{V} \frac{1}{\alpha^2 + K^2}.$$

Problem 3.16. Derive the Green's function $G_k(\vec{x}, \vec{x'})$ Equation (3.174) from Equation (3.172).

Solution 3.16. The contour integration of Equation (3.172) is outlined in the text. At the pole $k' = +k$ the upper semicircle contour gives $(2\pi i)$ times the residue $e^{ik|\vec{x} - \vec{x'}|}/2$, and at the pole $k' = -k$ the lower semicircle gives $-(2\pi i)$ times the residue $-e^{ik|\vec{x} - \vec{x'}|}/2$.

The results are the same, so the two parts add to give the Green's function (3.174).

Problem 3.17. Obtain Equation (3.183):

$$[\Phi(x), \Phi(y)] = \frac{1}{(2\pi)^3} \int \frac{d^3k}{2\omega} \{e^{-ik\cdot(x-y)} - e^{ik\cdot(x-y)}\}. \qquad (3.287)$$

Solution 3.17. The commutator of the field at two space-time points is:

$$[\Phi(x), \Phi(y)] = \frac{1}{(2\pi)^6} \int \frac{d^3k}{2\omega} \int \frac{d^3k'}{2\omega'}$$

$$\times \left\{ \left[a^\dagger(\vec{k}), a(\vec{k}') \right] e^{i(k\cdot x - k'\cdot y)} \right.$$

$$\left. + \left[a(\vec{k}), a^\dagger(\vec{k}') \right] e^{i(k'\cdot y - k\cdot x)} \right\};$$

$$\left[a(\vec{k}), a^\dagger(\vec{k}') \right] = (2\pi)^3 2\omega\delta(\vec{k} - \vec{k}').$$

Integration over $d^3k'\delta(\vec{k} - \vec{k}')$ gives Equation (3.183).

Problem 3.18. Verify Equation (3.197) for the photon propagator.

Solution 3.18. Take the Fourier expansion of the vector potential, Equation (3.110):

$$A_\mu(x) = \frac{1}{(2\pi)^3} \int \frac{d^3k}{2\omega} \left\{ a_\mu(\vec{k})e^{-ik\cdot x} + a_\mu^\dagger(\vec{k})e^{ik\cdot x} \right\}.$$

With the commutation relations:

$$\left[a_\mu(\vec{k}), a_\nu(\vec{k}') \right] = \left[a_\mu^\dagger(\vec{k}), a_\nu^\dagger(\vec{k}') \right] = 0;$$

$$\left[a_\mu(\vec{k}), a_\nu^\dagger(\vec{k}') \right] = -g_{\mu\nu}(2\pi)^3 2\omega\delta(\vec{k} - \vec{k}').$$

Then the commutator is:

$$[A_\mu(x), A_\nu(y)] = \frac{g_{\mu\nu}}{(2\pi)^3} \int \frac{d^3k}{2\omega} \left\{ e^{ik\cdot(x-y)} - e^{-ik\cdot(x-y)} \right\}.$$

Equation (3.197) is the photon propagator in momentum space:

$$\langle 0| T A_\mu(x) A_\nu(y) |0\rangle = \int \frac{d^4k}{(2\pi)^4} e^{-ik\cdot(x-y)} \left\{ \frac{-ig_{\mu\nu}}{(k^2 + i\epsilon)} \right\}.$$

The equivalence of this formula and the commutator is described in the text, by doing contour integration in $d\omega$ complex space on Equation (3.197) to reduce $d^4k = d^3k d\omega$ to $d^3k/(2\omega)$.

Problem 3.19. Show that Equation (3.204) is gauge invariant with $\psi \to \psi e^{ie\chi}$.

Solution 3.19. In Equation (3.204) the free field Lagrangian $-F_{\mu\nu}F^{\mu\nu}/4$ is gauge invariant, so we only need to look at the Dirac spinor terms:

$$A_\mu \to A_\mu + \partial_\mu\chi; \quad \psi \to \psi e^{ie\chi}; \quad i\bar{\psi}\gamma^\mu\partial_\mu\psi \to i\bar{\psi}\gamma^\mu\partial_\mu\psi - e(\partial_\mu\chi)\bar{\psi}\gamma^\mu\psi;$$

and the current term becomes:

$$eA_\mu\bar{\psi}\gamma^\mu\psi \to eA_\mu\bar{\psi}\gamma^\mu\psi + e(\partial_\mu\chi)\bar{\psi}\gamma^\mu\psi;$$

and the extra terms proportional to $\partial_\mu\chi$ cancel.

Problem 3.20. Obtain Equation (3.223) for the Dirac Hamiltonian.

Solution 3.20. Dirac Hamiltonian. Substitute the wave functions Equations (3.216) and (3.217) into Equation (3.215):

$$H = \frac{1}{(2\pi)^6} \int d^3x \int \frac{d^3p}{2E} \int \frac{d^3p'}{2E'} \sum_{s,s'}$$

$$\times \left\{ a_{s'}^\dagger(\vec{p}\,')\bar{u}_{s'}(p')e^{ip'\cdot x} + b_{s'}(\vec{p}\,')\bar{v}_{s'}(p')e^{-ip'\cdot x} \right\}$$

$$\times (-i\vec{\gamma}\cdot\vec{\nabla} + m)\left\{ a_s(\vec{p})u_s(p)e^{-ip\cdot x} + b_s^\dagger(\vec{p})v_s(p)e^{ip\cdot x} \right\}.$$

The product gives four terms:

$$a_{s'}^\dagger(\vec{p}\,')a_s(\vec{p})\bar{u}_{s'}(p')(\vec{\gamma}\cdot\vec{p} + m)u_s(p)e^{i(p'-p)\cdot x};$$

$$a_{s'}^\dagger(\vec{p}\,')b_s^\dagger(\vec{p})\bar{u}_{s'}(p')(-\vec{\gamma}\cdot\vec{p} + m)v_s(p)e^{i(p'+p)\cdot x};$$

$$b_{s'}(\vec{p}\,')a_s(\vec{p})\bar{v}_{s'}(p')(\vec{\gamma}\cdot\vec{p} + m)u_s(p)e^{-i(p'+p)\cdot x};$$

$$b_{s'}(\vec{p}\,')b_s^\dagger(\vec{p})\bar{v}_{s'}(p')(-\vec{\gamma}\cdot\vec{p} + m)v_s(p)e^{i(p-p')\cdot x};$$

The gradient operator $-i\vec{\nabla}$ is replaced by $\pm\vec{p}$, depending on the phase of the term in ψ. Now the space integral $1/(2\pi)^3 \int d^3x$ gives $\delta(\vec{p}-\vec{p}\,')$ for the first and fourth terms, and $\delta(\vec{p}+\vec{p}\,')$ for the two middle terms.

After integration over d^3p' the two middle terms have $\vec{p}\,' = -\vec{p}$, and the first and fourth terms have $\vec{p}\,' = \vec{p}$. Then as described in the text $(\not{p} - m)u(p) = 0$ and $(\not{p} + m)v(p) = 0$ can be used to replace $(\vec{\gamma} \cdot \vec{p} + m)u_s(p) = \gamma^0 E u_s(p)$ and $(-\vec{\gamma} \cdot \vec{p} + m)v_s(p) = -\gamma^0 E v_s(p)$. γ^0 reduces $\bar{u} \to u^\dagger$ and $\bar{v} \to v^\dagger$. Orthogonality of the spinors kills the two middle terms (you might want to check that Equation (3.221) is correct). We are left with:

$$H = \frac{1}{(2\pi)^3} \int \frac{d^3p}{4E} \sum_{s,s'} (a^\dagger_{s'}(\vec{p}) a_s(\vec{p}) u^\dagger_{s'}(p) u_s(p)$$

$$- b_{s'}(\vec{p}) b^\dagger_s(\vec{p}) v^\dagger_{s'}(p) v_s(p)).$$

Orthogonality and normalization of the spinors gives Equation (3.222), and normal ordering gives (3.223).

Chapter 4

Problem 4.1. Show the equivalence of various expressions for the flux in the cross-section formula Equations (4.11) and (4.12).

Solution 4.1. Drop the four and square both sides of Equation (4.11):

$$E_2^2|\vec{p}_1|^2 + E_1^2|\vec{p}_2|^2 + 2E_1E_2|\vec{p}_1||\vec{p}_2| = (p_1 \cdot p_2)^2 - m_1^2m_2^2.$$

This is true provided that $(p_1 \cdot p_2) = E_1E_2 + |\vec{p}_1||\vec{p}_2|$. Equation (4.12) squared is:

$$4((p_1 \cdot p_2)^2 - m_1^2m_2^2) = s^2 + m_1^4 + m_2^4 - 2sm_1^2 - 2sm_2^2 - 2m_1^2m_2^2.$$

Expand the right-hand side using $s = (p_1 + p_2)^2$:

$$s^2 + m_1^4 + m_2^4 - 2sm_1^2 - 2sm_2^2 - 2m_1^2m_2^2$$
$$= (m_1^2 + m_2^2)^2 + 4(p_1 \cdot p_2)^2 + 4p_1 \cdot p_2(m_1^2 + m_2^2) - 2(m_1^2 + m_2^2)^2$$
$$- 4p_1 \cdot p_2(m_1^2 + m_2^2) - 2m_1^2m_2^2 + m_1^4 + m_2^4$$
$$= -4m_1^2m_2^2 + 4(p_1 \cdot p_2)^2.$$

Problem 4.2. Derive Equation (4.25) starting with (4.19)–(4.21). Explain how the machinery works.

Solution 4.2. This problem is intended to oblige the reader to work through the reduction of the field theoretic calculation of the T matrix element for $e^-e^+ \to \mu^-\mu^+$. The field operators $a_s(\vec{p})$ and $b_s^\dagger(\vec{p})$ do not appear in the Feynman rules, and are almost never used after this derivation, so it is reasonable to go through these steps carefully at least once. The S matrix is Equation (4.19):

$$S = U_I(\infty, -\infty) = T\left\{-\frac{1}{2!}\int_{-\infty}^{\infty} dt H_I(t) \int_{-\infty}^{\infty} dt' H_I(t')\right\};$$

where

$$H_I(t) = -e\int d^3x : \bar\psi(\vec{x},t)\gamma^\mu\psi(\vec{x},t) : A_\mu(\vec{x},t).$$

The electron/positron Fourier analyzed wave functions are:

$$\psi(x) = \frac{1}{(2\pi)^3}\int \frac{d^3p}{2E}\sum_{s=1,2}\left\{a_s(\vec{p})u_s(p)e^{-ip\cdot x} + b_s^\dagger(\vec{p})v_s(p)e^{ip\cdot x}\right\};$$

and

$$\bar\psi(x) = \frac{1}{(2\pi)^3}\int \frac{d^3p'}{2E'}\sum_{s'=1,2}\left\{a_{s'}^\dagger(\vec{p}\,')\bar u_{s'}(p')e^{ip'\cdot x}\right.$$
$$\left. + b_{s'}(\vec{p}\,')\bar v_{s'}(p')e^{-ip'\cdot x}\right\}.$$

The muon wave function has the same form. Following the text we write $\psi_m(x')$, and the operators $c_r(\vec{q})$ and $d_r(\vec{q})$.

$$\psi_m(x') = \frac{1}{(2\pi)^3}\int \frac{d^3q}{2E_q}\sum_{r=1,2}\left\{c_r(\vec{q})u_r^m(q)e^{-iq\cdot x'} + d_r^\dagger(\vec{q})v_r^m(q)e^{iq\cdot x'}\right\};$$

and

$$\bar\psi_m(x') = \frac{1}{(2\pi)^3}\int \frac{d^3q'}{2E_q'}\sum_{r'=1,2}\left\{c_{r'}^\dagger(\vec{q}\,')\bar u_{r'}^m(q')e^{iq'\cdot x'}\right.$$
$$\left. + d_{r'}(\vec{q}\,')\bar v_{r'}^m(q')e^{-iq'\cdot x'}\right\}.$$

Look at the electron/positron Hamiltonian:

$$\int dt H_I(t) = -\frac{e}{(2\pi)^6}\int d^4x\int \frac{d^3p}{2E}\int \frac{d^3p'}{2E'}\sum_{s=1,2}\sum_{s'=1,2} : (a_{s'}^\dagger(\vec{p}\,')\bar u_{s'}(p')$$
$$\times e^{ip'\cdot x} + b_{s'}(\vec{p}\,')\bar v_{s'}(p')e^{-ip'\cdot x})\gamma^\mu(a_s(\vec{p})u_s(p)e^{-ip\cdot x}$$
$$+ b_s^\dagger(\vec{p})v_s(p)e^{ip\cdot x}) : A_\mu(x).$$

There are four terms:

$$: (a_{s'}^\dagger(\vec{p}\,')a_s(\vec{p})\bar{u}_{s'}(p')\gamma^\mu u_s(p)e^{i(p'-p)\cdot x}$$
$$+ a_{s'}^\dagger(\vec{p}\,')b_s^\dagger(\vec{p})\bar{u}_{s'}(p')\gamma^\mu v_s(p)e^{i(p'+p)\cdot x}$$
$$+ b_{s'}(\vec{p}\,')a_s(\vec{p})\bar{v}_{s'}(p')\gamma^\mu u_s(p)e^{-i(p'+p)\cdot x}$$
$$+ b_{s'}(\vec{p}\,')b_s^\dagger(\vec{p})\bar{v}_{s'}(p')\gamma^\mu v_s(p)e^{-i(p'-p)\cdot x}) : A_\mu(x)$$

The field theoretic expression contains four different processes. Look at the spinor currents to sort it out. The first is e^-e^- scattering, and the fourth is e^+e^+ scattering. The second and third are pair creation and pair annihilation respectively. It is the third term that is projected out by the creation operators on the vacuum: $a_{s_1}^\dagger(\vec{p}_1)b_{s_2}^\dagger(\vec{p}_2)|0\rangle$. The muon Hamiltonian has the same form, but the final state requires operators $\langle 0| a_{s_3}(\vec{p}_3)b_{s4}(\vec{p_4})$ that projects out the second term. Hence it is the two terms in the middle that contribute to $e^-e^+ \rightarrow \mu^-\mu^+$. As outlined in the text, there are four anti-commutators that give four spin Kronecker deltas and four vector momentum delta functions that couple discreet and continuum variables. The vacuum states collapse on the vector potentials. As there are no real photons in the problem, there are no discreet creation and destruction operators on the vacuum for the vector field A_μ. The photon propagator remains. It is important for this problem that the reader writes down all of the components and steps for himself/herself, in order to understand how the machinery of quantum field theory, that is the relativistic version of the quantum theory, and that contains the creation and annihilation of particles as done by nature, leads to the formulas of the Feynman rules.

Problem 4.3. Derive the formula for the T matrix element, Equation (4.28) from Equation (4.25).

Solution 4.3. This problem is a continuation of the preceding one, with the same objective, namely to go through the steps required to obtain the S matrix element for one's self. Evaluate the momentum integrals to take care of the δ functions in Equation (4.25) and substitute the photon propagator for the time ordered product to obtain

Equation (4.26):

$$\langle f | S | i \rangle = -e^2 \int d^4x \int d^4x' \bar{v}_{s_2}(p_2) \gamma^\mu u_{s_1}(p_1) \bar{u}_{s_3}^m(p_3) \gamma^\nu v_{s_4}^m(p_4)$$

$$\times e^{-i(p_1+p_2)\cdot x} e^{i(p_3+p_4)\cdot x'} \int \frac{d^4k}{(2\pi)^4} e^{-ik\cdot(x-x')} \left\{ \frac{-ig_{\mu,\nu}}{k^2 + i\epsilon} \right\}.$$

Now note that the four-dimensional space integrals are over coordinates that appear only in exponentials leading to two four momentum conservation delta functions.

$$\frac{1}{(2\pi)^4} \int d^4x\, e^{-i(p_1+p_2+k)\cdot x} = \delta^4(p_1 + p_2 + k); \quad \text{and}$$

$$\frac{1}{(2\pi)^4} \int d^4x'\, e^{i(p_3+p_4+k)\cdot x'} = \delta^4(p_3 + p_4 + k).$$

These two δ functions can be rewritten as $\delta^4(p_1 + p_2 - (p_3 + p_4))$ and $\delta^4(p_1 + p_2 + k)$. The first is overall momentum conservation, and the second takes care of k in the denominator of the propagator, giving Equation (4.28).

Problem 4.4. Derive Equation (4.33), showing how the sum over spins becomes a trace.

Solution 4.4. Start with Equation (4.30):

$$\frac{1}{4} \sum_{s_1,s_2} \bar{u}_{s_1}(p_1) \gamma^\nu v_{s_2}(p_2) \bar{v}_{s_2}(p_2) \gamma^\mu u_{s_1}(p_1)$$

$$= \frac{1}{4} \sum_{\alpha,\beta,\rho,\sigma} \sum_{\text{spins}} \bar{u}_\alpha(p_1) \gamma^\nu_{\alpha,\beta} v_\beta(p_2) \bar{v}_\rho(p_2) \gamma^\mu_{\rho,\sigma} u_\sigma(p_1).$$

Components can be shifted around to give

$$= \frac{1}{4} \sum_{\alpha,\beta,\rho,\sigma} \sum_{\text{spins}} u_\sigma(p_1) \bar{u}_\alpha(p_1) \gamma^\nu_{\alpha,\beta} v_\beta(p_2) \bar{v}_\rho(p_2) \gamma^\mu_{\rho,\sigma}.$$

Now use completeness:

$$\sum_{\text{spins}} u_\sigma(p_1) \bar{u}_\alpha(p_1) = (\not{p}_1)_{\sigma,\alpha} + m\delta_{\sigma,\alpha};$$

and

$$\sum_{\text{spins}} v_\beta(p_2)\bar{v}_\rho(p_2) = (\not{p}_2)_{\beta,\rho} - m\delta_{\beta,\rho}.$$

The spin sums over the spinors are now gone, leaving:

$$\frac{1}{4} \sum_{\alpha,\beta,\rho,\sigma} ((\not{p}_1)_{\sigma,\alpha} + m\delta_{\sigma,\alpha})\gamma^\nu_{\alpha,\beta}((\not{p}_2)_{\beta,\rho} - m\delta_{\beta,\rho})\gamma^\mu_{\rho,\sigma}$$

$$= \frac{1}{4}Tr((\not{p}_1 + m)\gamma^\nu(\not{p}_2 - m)\gamma^\mu).$$

Problem 4.5. Prove the identity $\gamma^\mu \not{a}\not{b}\gamma_\mu = 4a \cdot b$. From the definition of γ^5 justify the trace Equation (4.48), by choosing an order for μ, ν, ρ, σ and evaluating the product.

Solution 4.5. Proof of $\gamma^\mu \not{a}\not{b}\gamma_\mu = 4a \cdot b$.

$$\gamma^\mu \not{a}\not{b}\gamma_\mu = (2g^{\mu\rho} - \gamma^\rho\gamma^\mu)\gamma^\sigma \gamma_\mu a_\rho b_\sigma$$

$$= 2\not{b}\not{a} - \gamma^\rho(2g^{\mu\sigma} - \gamma^\sigma\gamma^\mu)\gamma_\mu a_\rho b_\sigma$$

$$= 2(\not{a}\not{b} + \not{b}\not{a}) = 4a \cdot b.$$

Let $\mu\nu\rho\sigma = 0123$ in Equation (4.48). Then

$$Tr(\gamma^0\gamma^1\gamma^2\gamma^3\gamma^5) = -4i; \quad \text{with} \quad \gamma^5 = i\gamma^0\gamma^1\gamma^2\gamma^3.$$

Use the anticommutation rules and $(\gamma^0)^2 = 1, (\gamma^j)^2 = -1$:

$$i\gamma^0\gamma^1\gamma^2\gamma^3\gamma^0\gamma^1\gamma^2\gamma^3 = -i\gamma^1\gamma^2\gamma^3\gamma^1\gamma^2\gamma^3$$

$$= +i\gamma^2\gamma^3\gamma^2\gamma^3 = -i; Tr(-i) = -4i.$$

Problem 4.6. Write down the muon tensor $M_{\mu\nu}$ and contract it with the electron tensor Equation (4.49), to obtain Equation (4.50).

Solution 4.6. Muon tensor:

$$M_{\mu\nu} = Tr(\not{p}_3\gamma_\nu\not{p}_4\gamma_\mu - m_\mu^2\gamma_\nu\gamma_\mu)$$

$$= 4(p_{3\nu}p_{4\mu} + p_{4\nu}p_{3\mu} - (p_3 \cdot p_4 + m_\mu^2)g_{\mu\nu}).$$

Too many μ's! Contract with the electron tensor Equation (4.49):

$$E^{\mu\nu}M_{\mu\nu} = 16(p_1^\mu p_2^\nu + p_2^\mu p_1^\nu - g^{\mu\nu}(p_1 \cdot p_2 + m_e^2))$$
$$\times (p_{3\nu}p_{4\mu} + p_{4\nu}p_{3\mu} - (p_3 \cdot p_4 + m_\mu^2)g_{\mu\nu})$$
$$= 16(2p_1 \cdot p_4 p_2 \cdot p_3 + 2p_1 \cdot p_3 p_2 \cdot p_4 - 2p_1 \cdot p_2 p_3 \cdot p_4$$
$$- 2p_1 \cdot p_2 p_3 \cdot p_4 - 2p_1 \cdot p_2 m_\mu^2$$
$$+ 4p_1 \cdot p_2 p_3 \cdot p_4 + 4p_1 \cdot p_2 m_\mu^2)$$
$$= 32(p_1 \cdot p_4 p_2 \cdot p_3 + p_1 \cdot p_3 p_2 \cdot p_4 + p_1 \cdot p_2 m_\mu^2).$$

We dropped the m_e^2 term.

Problem 4.7. Verify the high energy limit formulas in the center of mass, Equations (4.52)–(4.54).

Solution 4.7. HE limit in the center of mass, where all masses can be neglected, and $E_1 = E_2 = E_3 = E_4$. Four momentum conservation is $p_1 + p_2 = p_3 + p_4$. The Mandelstam variables are:

$$s = (p_1 + p_2)^2 = 2p_1 \cdot p_2 = 2(E_1 E_2 - \vec{p}_1 \cdot \vec{p}_2) = 4E^2;$$
$$t = (p_1 - p_3)^2 = -2p_1 \cdot p_3 = -2E^2(1 - \cos\theta) = -s(1 - \cos\theta)/2;$$
$$u = (p_1 - p_4)^2 = -s(1 + \cos\theta)/2.$$

Here the polar angle θ is between $\vec{p}_1$ and $\vec{p}_3$ in the center of mass.

Problem 4.8. Work out the explicit spinor matrix elements Equations (4.50)–(4.61), and obtain the electron tensor

$$E^{\mu,\nu} = \frac{1}{4} \sum_{\text{spins}} \bar{v}^{s_2} \gamma^\mu u^{s_1} \bar{u}^{s_1} \gamma^\nu v^{s_2} = 2E^2(\delta_{\mu,1}\delta_{\nu,1} + \delta_{\mu,2}\delta_{\nu,2}).$$

Obtain the same tensor from Equation (4.49) for $p_1^0 = p_2^0 = E$, and $p_1^3 = -p_2^3 = E$.

Solution 4.8. This is an exercise with spinors and matrices. For the electron-positron initial state, let the electron four momentum $p_1 = (E, 0, 0, E)$ and the positron $p_2 = (E, 0, 0, -E)$. The electron tensor is the product of two currents:

$$E^{\mu,\nu} = \frac{1}{4} \sum_{\text{spins}} \bar{v}^{s_2} \gamma^\mu u^{s_1} \bar{u}^{s_1} \gamma^\nu v^{s_2};$$

and we are asked to show that this equals:

$$= 2E^2 (\delta_{\mu,1}\delta_{\nu,1} + \delta_{\mu,2}\delta_{\nu,2}).$$

Work with $\vec{\alpha}$, β, $u^{(s)\dagger}(p_1)$, and $v^{(s')}(p_2)$. The other matrix elements are hermitian conjugates.

$$u^{(1)\dagger}(p_1) = \sqrt{E}\begin{pmatrix} 1 & 0 & 1 & 0 \end{pmatrix}; \quad u^{(2)\dagger}(p_1) = \sqrt{E}\begin{pmatrix} 0 & 1 & 0 & -1 \end{pmatrix};$$

$$v^{(1)}(p_2) = \sqrt{E}\begin{pmatrix} 0 \\ 1 \\ 0 \\ 1 \end{pmatrix}; \quad v^{(2)}(p_2) = \sqrt{E}\begin{pmatrix} -1 \\ 0 \\ 1 \\ 0 \end{pmatrix}.$$

The operators are:

$$\beta = \begin{pmatrix} 1 & 0 & 0 & 0 \\ 0 & 1 & 0 & 0 \\ 0 & 0 & -1 & 0 \\ 0 & 0 & 0 & -1 \end{pmatrix}; \quad \alpha^1 = \begin{pmatrix} 0 & 0 & 0 & 1 \\ 0 & 0 & 1 & 0 \\ 0 & 1 & 0 & 0 \\ 1 & 0 & 0 & 0 \end{pmatrix};$$

$$\alpha^2 = \begin{pmatrix} 0 & 0 & 0 & -i \\ 0 & 0 & i & 0 \\ 0 & -i & 0 & 0 \\ i & 0 & 0 & 0 \end{pmatrix}; \quad \alpha^3 = \begin{pmatrix} 0 & 0 & 1 & 0 \\ 0 & 0 & 0 & -1 \\ 1 & 0 & 0 & 0 \\ 0 & -1 & 0 & 0 \end{pmatrix}.$$

Now multiply it out:

$$u^\dagger v = \begin{pmatrix} 1,1 & 1,2 & 2,1 & 2,2 \\ 0 & 0 & 0 & 0 \end{pmatrix};$$

$$u^\dagger \alpha^1 v = \begin{pmatrix} 1,1 & 1,2 & 2,1 & 2,2 \\ 2E & 0 & 0 & 2E \end{pmatrix}$$

$$u^\dagger \alpha^2 v = \begin{pmatrix} 1,1 & 1,2 & 2,1 & 2,2 \\ -2iE & 0 & 0 & 2iE \end{pmatrix};$$

$$u^\dagger \alpha^3 v = \begin{pmatrix} 1,1 & 1,2 & 2,1 & 2,2 \\ 0 & 0 & 0 & 0 \end{pmatrix}.$$

(1)

The hermetian conjugates are the same except for α^2, where $(1,1)$ and $(2,2)$ interchange. In the electron tensor above the e^- spins are the same in the two currents, as are the e^+ spins. There are no terms with either $\mu = 0$, $\mu = 3$, $\nu = 0$, or $\nu = 3$. Sum spins for each combination of μ and ν:

$$E^{1,1} = \frac{1}{4}(4E^2 + 4E^2) = 2E^2; \quad E^{1,2} = \frac{1}{4}(-4iE^2 + 4iE^2) = 0;$$

$$E^{2,1} = \frac{1}{4}(4iE^2 - 4iE^2) = 0; \quad E^{2,2} = \frac{1}{4}(4E^2 + 4E^2) = 2E^2.$$

This gives the desired result.

Problem 4.9. Using the same techniques derive the muon tensor Equation (4.66), and the contraction of the two $E^{\mu\nu} M_{\mu\nu}$ Equation (4.67).

Solution 4.9. The angular dependence of the final state is contained in the spinors. For the μ^- $p_3 = (E, E \sin\theta, 0, E \cos\theta)$, and for the μ^+, $p_4 = (E, -E \sin\theta, 0, -E \cos\theta)$. The spinors are:

$$u^{(1)}(p_3) = \sqrt{E} \begin{pmatrix} 1 \\ 0 \\ \cos\theta \\ \sin\theta \end{pmatrix}; \quad u^{(2)}(p_3) = \sqrt{E} \begin{pmatrix} 0 \\ 1 \\ \sin\theta \\ -\cos\theta \end{pmatrix};$$

and

$$v^{(1)\dagger}(p_4) = \sqrt{E}\left(-\sin\theta \quad \cos\theta \quad 0 \quad 1\right);$$

$$v^{(2)\dagger}(p_4) = \sqrt{E}\left(-\cos\theta \quad -\sin\theta \quad 1 \quad 0\right).$$

You only need α^1 and α^2 since these are the components of the electron tensor. The calculation is straightforward.

Problem 4.10. Obtain Equation (4.76) for $e^-\mu^-$ scattering from Equation (4.50) for $e^-e^+ \to \mu^-\mu^+$.

Solution 4.10. Refer to Figure 1. This clearly shows how the four momenta are shuffled, and explains the mapping of the Mandelstam variables: $s \to t'; t \to u'$; and $u \to s'$. Hence Equation (4.55) becomes Equation (4.78).

e⁻e⁺->μ⁻μ⁺ and e⁻μ⁻->e⁻μ⁻

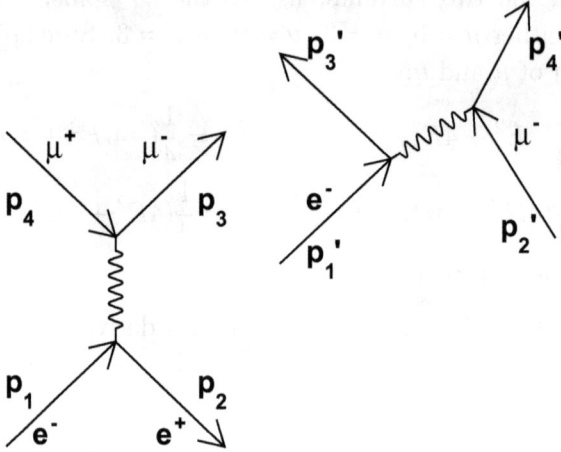

Figure 1: Feynman diagrams for s-channel annihilation $e^- e^+ \rightarrow \mu^- \mu^+$ and t-channel $e^- \mu^-$ elastic scattering. The incident electron $p_1 = p_1'$. p_2 is the incident positron that becomes the final state electron, i.e. $p_2 \rightarrow -p_3'$. The final state μ^+ becomes the initial state μ^-, so $p_4 \rightarrow -p_2'$. Finally $p_3 \rightarrow p_4'$, the final state μ^- remains, but switches labels because in the annihilation diagram p_1 in the initial state flows to the negative muon p_3 in the final state.

Problem 4.11. Draw the diagrams for Bhabha scattering, and use the Feynman rules to construct the amplitudes from the diagrams. Verify the trace calculations for the three terms $|A(s)|^2$, $|A(t)|^2$, and $2A^\dagger(t)A(s)$.

Solution 4.11. Bhabha scattering $e^- e^+ \rightarrow e^- e^+$ has two amplitudes called $A(s)$ for the annihilation diagram and $A(t)$ for the scattering diagram. The analysis is well described in the text. The reader is asked to repeat the steps for practice.

$$\frac{1}{4} \sum_{\text{spins}} |A(s)|^2 \text{ and } \frac{1}{4} \sum_{\text{spins}} |A(t)|^2$$

both factor into two traces of four γ matrices each, giving the formulas in terms of the Mandelstam variables in Equation (4.83). The interference term does not factor, but can be reduced using the "non-trace" theorems Equations (4.41) and (4.42).

$$\frac{1}{2}\sum_{\text{spins}} A^\dagger(s)A(t) = \frac{e^4}{2st}Tr(\not{p}_4\gamma_\mu\not{p}_3\gamma^\lambda\not{p}_1\gamma^\mu\not{p}_2\gamma_\lambda).$$

Now use Equation (4.42) to get rid of the $\gamma^\mu\gamma_\mu$ pair:

$$\gamma_\mu\not{p}_3\gamma^\lambda\not{p}_1\gamma^\mu = -2\not{p}_1\gamma^\lambda\not{p}_3;\ \text{leaving} -\frac{e^4}{st}Tr(\not{p}_4\not{p}_1\gamma^\lambda\not{p}_3\not{p}_2\gamma_\lambda).$$

Then Equation (4.41) gives the final result:

$$\frac{1}{2}\sum_{\text{spins}} A^\dagger(s)A(t) = -4\frac{e^4u^2}{st}. \tag{2}$$

This is Equation (4.87) with a minus sign.

Problem 4.12. Repeat the steps outlined in Problem 11 for Møller scattering.

Solution 4.12. Møller scattering $e^-e^- \to e^-e^-$ also involves two amplitudes, called $A(t)$ and $A(u)$, the latter being the crossed diagram. Again the two amplitudes are subtracted. The analysis is described in the text, and follows that of Bhabha scattering very closely.

Problem 4.13. Verify Equation (4.108) for the spin projection operator Λ_+, and show that $\Lambda_+u^1(p) = u^1(p)$, and $\Lambda_+u^2(p) = 0$.

Solution 4.13. $\Lambda_+ = (1 + \gamma^5\not{s})/2$ with the spin four vector $s = (p/m, 0, 0, E/m)$. So:

$$\not{s} = \gamma^0 p/m - \gamma^3 E/m = \begin{pmatrix} p/m & 0 & -E/m & 0 \\ 0 & p/m & 0 & E/m \\ E/m & 0 & -p/m & 0 \\ 0 & -E/m & 0 & -p/m \end{pmatrix};$$

and

$$\gamma^5 = \begin{pmatrix} 0 & 0 & 1 & 0 \\ 0 & 0 & 0 & 1 \\ 1 & 0 & 0 & 0 \\ 0 & 1 & 0 & 0 \end{pmatrix} \quad \text{so}$$

$$\gamma^5 \displaystyle{\not}p = \begin{pmatrix} E/m & 0 & -p/m & 0 \\ 0 & -E/m & 0 & -p/m \\ p/m & 0 & -E/m & 0 \\ 0 & p/m & 0 & E/m \end{pmatrix}.$$

Adding the unit matrix and dividing by two gives Equation (4.108). Operating on the spinor $u^{(1)}(p)$ gives:

$$\frac{1+\gamma^5 \displaystyle{\not}p}{2} u^{(1)}(p) = \frac{\sqrt{E+m}}{2} \begin{pmatrix} 1 + E/m - p^2/(m(E+m)) \\ 0 \\ p/m + (1 - E/m)p/(E+m) \\ 0 \end{pmatrix}$$

$$= u^{(1)}(p).$$

Equation (2.122) gives $u^{(2)}(p)$:

$$(1+\gamma^5 \displaystyle{\not}p)u^{(2)}(p) = \sqrt{E+m} \begin{pmatrix} 0 \\ 1 - E/m + p^2/(m(E+m)) \\ 0 \\ p/m - (1 + E/m)p/(E+m) \end{pmatrix} = 0.$$

Problem 4.14. Verify for $\alpha\mu\beta\nu = 0123$ that $\text{Tr}(\gamma^\alpha\gamma^\mu\gamma^\beta\gamma^\nu\gamma^5) = -4i$.

Solution 4.14. This is a repeat of Problem 4.5, clearly showing overzealous enthusiasm for γ^5 on the part of the author.

Problem 4.15. Derive the electron tensor $E_{RL}^{\mu\nu}$ of Equation (4.119), and the muon tensor $M_{\mu\nu}^{RL}$ of Equation (4.123).

Solution 4.15. Polarization in $e^- e^+ \to \mu^- \mu^+$. Insert the spin projection operator Λ_+ into the electron tensor, Equation (4.114).

$$E_{RL}^{\mu\nu} = \frac{1}{4} \sum_{\text{spins}} \bar{v}^{s_2} \gamma^\mu u^{s_1} \bar{u}^{s_1} \gamma^\nu v^{s_2} = \frac{1}{8}(\mathrm{Tr}(\not{p}_2 \gamma^\mu (\not{p}_1 + m)(1 + \gamma^5 \not{s}_1)\gamma^\nu)).$$

The electron mass is retained because it appears in the denominator of the spin vector s^μ.

$$E_{RL}^{\mu\nu} = \frac{1}{8}\mathrm{Tr}(\not{p}_2 \gamma^\mu \not{p}_1 \gamma^\nu) + \frac{m}{8}\mathrm{Tr}(\not{p}_2 \gamma^\mu \gamma^5 \not{s}_1 \gamma^\nu).$$

The first term is the unpolarized part, and the second term involves γ^5:

$$\frac{m}{8}\mathrm{Tr}(\not{p}_2 \gamma^\mu \gamma^5 \not{s}_1 \gamma^\nu) = \frac{m}{8}\mathrm{Tr}(\gamma^\alpha \gamma^\mu \gamma^5 \gamma^\beta \gamma^\nu)p_{2\alpha}s_{1\beta} = \frac{im}{2}\epsilon^{\alpha\mu\beta\nu}p_{2\alpha}s_{1\beta}.$$

The four vectors are: $p_1 = (E, 0, 0, E)$, $p_2 = (E, 0, 0, -E)$ and $s_1 = (E/m, 0, 0, E/m)$. p_2 and s_1 have only components 0 and 3, so μ and ν can be only 1 or 2, since ϵ requires all four indices to be different.

$$\frac{m}{8}\mathrm{Tr}(\not{p}_2 \gamma^\mu \gamma^5 \not{s}_1 \gamma^\nu) = \frac{im}{2}\left(\epsilon^{0\mu3\nu}\frac{E^2}{m} + \epsilon^{3\mu0\nu}\left(-\frac{E^2}{m}\right)\right) = iE^2\epsilon^{0\mu3\nu}.$$

Interchanging 0 and 3 on the second ϵ changes its sign, and the overall sign comes from $\epsilon^{0123} = -1$. Hence:

$$E_{RL}^{\mu\nu} = E^2(\delta_{\mu,1}\delta_{\nu,1} + \delta_{\mu,2}\delta_{\nu,2} + i(\delta_{\mu,1}\delta_{\nu,2} - \delta_{\mu,2}\delta_{\nu,1})).$$

The first part is unpolarized, and the second, imaginary part is the polarization. The normal calculation of the unpolarized part, dropping the electron mass, is:

$$E^{\mu\nu} = p_1^\mu p_2^\nu + p_1^\nu p_2^\mu - p_1 \cdot p_2 g^{\mu\nu}.$$

One might ask how it is that only $\mu = 1, \nu = 1$ and $\mu = 2, \nu = 2$ contribute, when the four momenta p_1 and p_2 have only components $\mu = 0$ and $\mu = 3$? The answer lies with the metric tensor. The reader is invited to verify that it works. Clue: $p_1 \cdot p_2 = 2E^2$. The muon spinors depend on the scattering angle θ, but the analysis is much the same. The four vectors are:

$$p_3 = (E, E\sin\theta, 0, E\cos\theta); \quad p_4 = (E, -E\sin\theta, 0, -E\cos\theta);$$

and in the relativistic limit the spin four vector is:

$$s_3 = (E/m_\mu, (E\sin\theta)/m_\mu, 0, (E\cos\theta)/m_\mu).$$

Problem 4.16. Contract the two tensors of Problem 4.15 to obtain the angular distribution for polarized leptons.

Solution 4.16. The RL muon tensor is

$$M_{\mu\nu}^{RL} = \frac{1}{2}Tr(\not{p}_4\gamma_\mu\not{p}_3\gamma_\nu) + \frac{m_\mu}{2}Tr(\not{p}_4\gamma_\mu\gamma^5\not{p}_3\gamma_\nu)$$

$$= 4E^2(\cos^2\theta\delta_{\mu,1}\delta_{\nu,1} + \delta_{\mu,2}\delta_{\nu,2} - i\cos\theta(\delta_{\mu,1}\delta_{\nu,2} - \delta_{\mu,2}\delta_{\nu,1})).$$

The contraction of the two tensors is:

$$RL \to RL = 4E^4(\cos^2\theta\delta_{\mu,1}\delta_{\nu,1} + \delta_{\mu,2}\delta_{\nu,2}$$

$$- i\cos\theta(\delta_{\mu,1}\delta_{\nu,2} - \delta_{\mu,2}\delta_{\nu,1}))$$

$$\times (\delta_{\mu,1}\delta_{\nu,1} + \delta_{\mu,2}\delta_{\nu,2} + i(\delta_{\mu,1}\delta_{\nu,2} - \delta_{\mu,2}\delta_{\nu,1}))$$

$$= 4E^2(\cos^2\theta + 1 + 2\cos\theta). \tag{3}$$

The angular dependence is $(1 + \cos\theta)^2$, which vanishes in the backward direction for the μ^-. θ is the polar angle between the incident e^- and the final state μ^-. If the μ^- is right handed, it cannot come out backwards and conserve angular momentum. The intermediate photon does not transmit the flow of charge, but it does transmit the spin.

Problem 4.17. Sharpen your pencil, and evaluate the traces found in $(1/4)\sum_{\text{spins}}|A|^2$ for Compton scattering, Equations (4.142) and (4.143).

Solution 4.17. The traces in Equation (4.142) are straightforwardly taken one at a time.

$$Tr(\not{p}'\gamma_\alpha\not{k}\gamma_\beta\not{p}\gamma^\beta\not{k}\gamma^\alpha) = -2Tr(\not{p}'\gamma_\alpha\not{k}\not{p}\not{k}\gamma^\alpha) = 4Tr(\not{p}'\not{k}\not{p}\not{k})$$

$$= 32(p' \cdot k)(p \cdot k).$$

$$2Tr(\not{p}'\gamma_\alpha\not{k}\not{p}\not{p}\gamma^\alpha) = -4Tr(\not{p}'\not{p}\not{p}\not{k}) = -16m_e^2(p' \cdot k).$$

$$2Tr(\not{p}'\gamma_\alpha\not{p}\not{p}\not{k}\gamma^\alpha) = -4Tr(\not{p}'\not{k}\not{p}\not{p}) = -16m_e^2(p' \cdot k).$$

$$4Tr(\not{p}'\gamma_\alpha p_\beta\not{p}\gamma^\alpha)p^\beta = 4m_e^2Tr(\not{p}'\gamma_\alpha\not{p}\gamma^\alpha) = -32m_e^2(p' \cdot p). \tag{4}$$

Problem 4.18. Sharpen it again, and evaluate the traces for the cross term $(1/2)\sum_{\text{spins}}A^\dagger B$, Equation (4.149).

Solution 4.18. Evaluation of the traces in Equation (4.148) to give Equation (4.149) for the cross term $\sum_{\text{spins}} A^\dagger B/2$ for Compton scattering is very similar to Problem 4.17. Evaluate each term, and then combine them.

Problem 4.19. Show that Equation (4.150) becomes the Klein–Nishina Formula Equation (4.151).

Solution 4.19. Equation (4.150) reads:

$$\frac{1}{4}\sum_{\text{spins}}(A^\dagger + B^\dagger)(A + B) = 2e^4\left\{\frac{k'\cdot p}{k\cdot p} + \frac{k\cdot p}{k'\cdot p} + 2m^2\left(\frac{1}{k\cdot p}\right.\right.$$

$$\left.\left. - \frac{1}{k'\cdot p}\right) + m^4\left(\frac{1}{k\cdot p} - \frac{1}{k'\cdot p}\right)^2\right\}.$$

Use $k\cdot p = m\omega$, $k'\cdot p = m\omega'$, and $1/\omega' - 1/\omega = (1 - \cos\theta)/m$:

$$\frac{1}{4}\sum_{\text{spins}}(A^\dagger + B^\dagger)(A + B)$$

$$= 2e^4\left\{\frac{\omega'}{\omega} + \frac{\omega}{\omega'} - 2(1 - \cos\theta) + (1 - \cos\theta)^2\right\}.$$

This equals Equation (4.151).

Problem 4.20. Perform the integrals for the unpolarized and polarized Compton cross-sections to obtain Equations (4.160) and (4.161).

Solution 4.20. The integrals for the Compton total cross-section are all elementary, as the text says. The unpolarized cross-section, apart from constants, is:

$$\int d\cos\theta\frac{\omega'^2}{\omega^2}\left(\frac{\omega}{\omega'} + \frac{\omega'}{\omega} - \sin^2\theta\right)$$

$$= \int dx\left\{\frac{1}{(1 + k(1 - x))^3} + \frac{1}{1 + k(1 - x)} - \frac{1 - x^2}{(1 + k(1 - x))^2}\right\};$$

where $k = \omega/m_e$, $x = \cos\theta$, and $1 + k(1 - x) = \omega/\omega'$. Let $y = 1 + k(1 - x)$, $dy = -kdx$, $x = 1 + (1 - y)/k$, limits $x = 1, y = 1$, and $x = -1, y = 1 + 2k$. The first term is:

$$\frac{1}{k} \int_1^{1+2k} \frac{dy}{y^3} = \frac{1}{k} \left(\frac{1}{2} - \frac{1}{2(1+2k)^2} \right).$$

The second term is:

$$\frac{1}{k} \int_1^{1+2k} \frac{dy}{y} = \frac{\ln(1+2k)}{k};$$

the third term is:

$$-\frac{1}{k} \int_1^{1+2k} \frac{dy}{y^2} = -\frac{2}{1+2k}.$$

The last term is:

$$\frac{1}{k} \int_1^{1+2k} \frac{(1+k)^2 - 2y(1+k) + y^2}{(ky)^2} dy$$

$$= \frac{2(1+k)^2}{k^2(1+2k)} - 2\frac{(1+k)\ln(1+2k)}{k^3} + \frac{2}{k^2}.$$

Assembling all of the parts gives Equation (4.160). The polarized cross-section, Equation (4.161), is a similar exercise.

Problem 4.21. Compton back scattering of laser light. Shine a $\lambda = 500\,\text{nm}$ laser (blue-green light) directly at an incident $50\,\text{GeV}$ electron beam. Calculate the energy of the back scattered γ rays in the laboratory. You may work either in the center of mass, with Lorentz transformations in and out, or stay in the laboratory. In either case you have to be careful in your approximations, because the electron mass cannot be ignored. The center of mass is moving in the electron direction. With laser photon momentum k and electron momentum p, the velocity of the center of mass is $v_c = (p - k)/(k + \sqrt{m^2 + p^2}) \approx 1 - 2k/p - m^2/(2p^2)$. For the given energies, k and $m^2/(2p)$ are about the same, roughly $2.5\,\text{eV}$. Obtain the energy of the back scattered γ ray in the laboratory:

$$k' = \frac{p}{1 + m^2/(4kp)}.$$

For our numbers, this amounts to $k' = 2p/3$, leaving the scattered electron (still traveling in its original direction) with $1/3$ of its original energy. You get $33\,\text{GeV}$ γ rays coming back at you! The laser and electron beams have to be very carefully aligned, and the resulting γ ray beam is not very intense, but it can be polarized by polarizing the original laser light, and has been used to study spin dependent effects in photoproduction.

Solution 4.21. I worked in the laboratory. It is important to realize that p, p', and k' are all about the same size, while m_e is small, and k is even smaller. Conservation of energy and momentum in the laboratory read:

$$k + \sqrt{p^2 + m_e^2} = k' + \sqrt{p'^2 + m_e^2}; \text{ and } p - k = k' + p'.$$

Eliminate p':

$$k + \sqrt{p^2 + m_e^2} = k' + \sqrt{((p - k') - k)^2 + m_e^2};$$

now expand both sides:

$$k + p + \frac{m_e^2}{2p} = k' + (p - k')\left(1 - \frac{1}{2}\left(\frac{2k}{p + k'} + \frac{k^2 + m_e^2}{(p - k')^2}\right)\right).$$

Combining terms:

$$\frac{p - k'}{k^2 + m_e^2} = \frac{p}{4pk + m_e^2}; \quad \text{and} \quad k' = p\left(1 - \frac{k^2 + m_e^2}{4pk + m_e^2}\right); \text{ giving}$$

$$k' = \frac{p}{1 + m_e^2/(4kp)}.$$

This approximation works because $k^2 \ll 4pk$. For 500 nm laser light, $k = 1240/500 = 2.5$ eV. $p = 5 \times 10^{10}$ eV, $m_e^2 = 25 \times 10^{10}$ eV^2, and $m_e^2/(4kp) = 1/2$, giving $k' = 2p/3 = 33\,\text{GeV}$.

Chapter 5

Problem 5.1. Verify Equations (5.10) and (5.11).

Solution 5.1. Start with Equations (5.7) and (5.8) The electron tensor is:

$$E^{\mu\lambda} = \text{Tr}(\not{k}'\gamma^\mu\not{k}\gamma^\lambda) = 4(k'^\mu k^\lambda + k'^\lambda k^\mu - (k' \cdot k)g^{\mu\lambda}).$$

For the muon tensor we retain the muon mass term:

$$M_{\mu\lambda} = \text{Tr}((\not{p}' + m_\mu)\gamma_\mu(\not{p} + m_\mu)\gamma_\lambda)$$

$$= 4(p'_\mu p_\lambda + p'_\lambda p_\mu + (m_\mu^2 - (p' \cdot p))g_{\mu\lambda}).$$

The contraction is:

$$E^{\mu\lambda}M_{\mu\lambda} = 32((k' \cdot p')(k \cdot p) + (k \cdot p')(k' \cdot p) - m_\mu^2(k' \cdot k)).$$

Then Equation (5.6) gives Equation (5.10):

$$\frac{1}{4}\sum_{spins} |\langle f| T |i\rangle|^2 = \overline{|\langle f| T |i\rangle|^2} = \frac{e^4}{4q^4}E^{\mu\lambda}M_{\mu\lambda} = \frac{8e^4}{q^4}((k' \cdot p')(k \cdot p)$$

$$+ (k \cdot p')(k' \cdot p) - m_\mu^2(k' \cdot k))$$

Then use $k \cdot k' = 2EE'\sin^2(\theta/2)$, $k \cdot p = k' \cdot p' = m_\mu E$, $k' \cdot p = k \cdot p' = m_\mu E'$ to obtain:

$$\frac{8e^4 m_\mu^2}{q^4}(E^2 + E'^2 - 2EE'\sin^2(\theta/2)).$$

This expression equals Equation (5.11) if:

$$2EE'(\cos^2(\theta/2) - \frac{q^2}{2m_\mu^2} \sin^2(\theta/2)) = E^2 + E'^2 - 2EE' \sin^2(\theta/2).$$

Use $-q^2 = 4EE' \sin^2(\theta/2) = 2m_\mu(E - E')$ to show that this equality is true. There are so many different kinematic relations that it is simpler to demonstrate the correctness of this form than it is to derive it.

Problem 5.2. Using the elastic scattering relation $-\frac{q^2}{2m\nu} = 1$, confirm the kinematic relation Equation (5.14):

$$\frac{E'}{E} = \frac{1}{1 + \frac{2E \sin^2(\theta/2)}{M}},$$

and obtain Equation (5.15) from Equation (5.12).

Solution 5.2. Equation (5.14) comes directly from the elastic kinematic formula $-q^2 = 2m_\mu(E - E') = 4EE' \sin^2(\theta/2)$, namely:

$$\frac{2EE'}{m_\mu} \sin^2(\theta/2) = E - E'; \quad \frac{E}{E'} = 1 + \frac{2E \sin^2(\theta/2)}{m_\mu}.$$

Equation (5.12) reads:

$$d\sigma = (2\pi)^4 \delta^4(p + k - p' - k') \frac{d^3 k'}{(2\pi)^3 2E'} \frac{d^3 p'}{(2\pi)^3 2p_0'} |\langle f| \mathrm{T} |i \rangle|^2 \frac{1}{4(mE)}.$$

To obtain Equation (5.15) one has to integrate over the δ function.

$$\delta^4(p + k - k' - p') = \delta(m_\mu + E - E' - p_0') \delta(\vec{k} - \vec{k}' - \vec{p}').$$

Integrate over $d^3 p'$ with the momentum δ function and replace the recoil energy $p_0' = \sqrt{m_\mu^2 + (\vec{k} - \vec{k}')^2}$. Using $|\vec{k}| = E$ and $|\vec{k}'| = E'$ the energy δ function becomes $\delta(m_\mu + E - E' - \sqrt{m_\mu^2 + E^2 + E'^2 - 2EE' \cos\theta})$, giving $f(E') = E' + \sqrt{m_\mu^2 + E^2 + E'^2 - 2EE' \cos\theta}$; and $df/dE' = (p_0' + E' - E\cos\theta)/p_0' = (m_\mu + E(1 - \cos\theta))/p_0'$. The inverse cancels the p_0' in the denominator and gives the factor $(E'/(m_\mu E))^2$ in Equation (5.13).

Problem 5.3. Derive the Mott cross-section Equation (5.17). Suggested procedure: put the electron mass back in the electron tensor $E^{\mu\lambda}$, and contract it with a tensor $M_{\mu\lambda}$ for which $p = p' = (M, 0, 0, 0)$. This should give $E^{\mu\lambda}M_{\mu\lambda} = 32M^2(E^2 + m_e^2 + k^2\cos\theta)$. Obtain $|\langle f|\,\mathrm{T}\,|i\rangle|^2$, averaging over initial spins (our heavy target has spin 1/2). Now modify Equation (5.12) for one outgoing particle in the final state, the target still being at rest as follows:

$$d\sigma = \frac{2\pi\delta(E - E')}{2M}\frac{d^3k'}{(2\pi)^3 2E'}|\langle f|\,\mathrm{T}\,|i\rangle|^2\frac{1}{4kM},\tag{5.101}$$

where $1/2M$ normalizes the spinor, the last factor is the flux, and $k = k'$ is the magnitude of the electron momentum. In the matrix element $E^2(1 + \frac{m_e^2}{E^2} + v^2\cos\theta) = 2E^2(1 - v^2\sin^2(\theta/2))$; so after some cancellation and integrating over the δ function, one obtains

$$\frac{d\sigma}{d\Omega} = \frac{\alpha^2 E^2}{4k^4\sin^4(\theta/2)}(1 - v^2\sin^2(\theta/2)).\tag{5.102}$$

As mentioned in the text, in the limit $v \to 0$, this is Rutherford's formula for scattering of α particles from gold, after multiplying by $4Z^2$, and for gold $Z = 79$. It is interesting that we started out with two spin 1/2 Dirac particles, with charges and magnetic moments, scattering via electric and magnetic forces, but the formula we wound up with in the non-relativistic limit is the same as for spinless α particles. In fact it is the same as Rutherford derived using classical mechanics. In the Mott form, the factor $(1 - v^2\sin^2(\theta/2))$ is a helicity rule for the electron spin, that becomes one in the NR limit.

Solution 5.3. Mott cross-section. Follow the suggestions in the problem.

$$E^{\mu\lambda} = 4(k'^\mu k^\lambda + k'^\lambda k^\mu + (m_e^2 - k'\cdot k)g^{\mu\lambda}); \; M_{\mu\lambda}$$
$$= 4(p'_\mu p_\lambda + p'^\lambda p_\mu + (M^2 - p'\cdot p)g_{\mu\lambda}).$$

Substituting the four vectors for a target at rest that remains at rest: $p^\mu = p'^\mu = (M, 0, 0, 0)$ gives $M_{\mu\lambda} = 8M^2\delta_{\mu,0}\delta_{\lambda,0}$. The contraction then is:

$$E^{\mu\lambda}M_{\mu\lambda} = 32M^2(EE' + E'E + m_e^2 - E'E + k'k\cos\theta)$$
$$= 32M^2(E^2 + m_e^2 + k^2\cos\theta).$$

For a heavy target $E' = E$ and $k' = k$, but we allow for the non-relativistic limit on the electron, so that E and k are not the same. The electron velocity is $v = k/E$. Using Equation (5.6):

$$\frac{1}{4}\sum_{\text{spins}}|\langle f|T|i\rangle|^2 = \frac{e^4}{4q^4} \times 32M^2(E^2 + m_e^2 + k^2\cos\theta).$$

The one-particle final state cross-section formula is given by Equation (5.101):

$$d\sigma = \frac{2\pi\delta(E-E')}{2M}\frac{d^3k'}{(2\pi)^3 2E'}\left\{\frac{8e^4M^2}{q^4}(E^2 + m_e^2 + k^2\cos\theta)\right\}\frac{1}{4kM};$$

$\int \delta(E-E')d^3k'/(2E') = \int \delta(E-E')k'E'dE'd\Omega/(2E') = k'd\Omega/2,$ giving:

$$\frac{d\sigma}{d\Omega} = \frac{e^4}{128\pi^2 k^4\sin^4(\theta/2)}(E^2 + m_e^2 + k^2\cos\theta)$$

$$= \frac{\alpha^2}{4k^4\sin^4(\theta/2)}E^2(1 - v^2\sin^2(\theta/2));$$

where we have used $\alpha = e^2/(4\pi)$, and $E^2 + m_e^2 + k^2\cos\theta = 2E^2(1 - v^2\sin^2(\theta/2))$, in agreement with Equation (5.102). To get Rutherford's non-relativistic formula, reduce $k^2 = E^2 - m_e^2 = (E-m_e)(E+m_e) \sim KE \times 2m_e$. So $k^4 = 4(KE)^2 m_e^2$, and $E^2 \sim m_e^2$ in the numerator; setting $v^2 \sim 0$ leaves:

$$\frac{d\sigma}{d\Omega} = \frac{\alpha^2}{16(KE)^2}\frac{1}{\sin^4(\theta/2)}.$$

This is Rutherford's formula in our units, for $ZZ' = 1$. To convert to barns, insert the particle KE in GeV, and use $1\text{ GeV}^{-2} = 0.389$ mb. For 5 MeV α particles on gold, $ZZ' = 158$, and the numerical constant is 1.3 barns, in agreement with Equation (3.162).

Problem 5.4. Evaluate the traces in Equation (5.24) to obtain $M^{\mu\lambda}$ in Equation (5.25). Then contract $M^{\mu\lambda}$ with the electron tensor $E_{\mu\lambda}$ to obtain Equation (5.27).

Solution 5.4. Equation (5.24) has a whole lot of traces, but they are all easy! Take them one at a time.

$$M^{\mu\lambda} = (F_1 + \kappa F_2)^2 \text{Tr}((\not{p}' + M)\gamma^\mu(\not{p} + M)\gamma^\lambda) - (F_1 + \kappa F_2)\frac{\kappa F_2}{2M}$$

$$\times \text{Tr}((\not{p}' + M)\gamma^\mu(\not{p} + M)(p + p')^\lambda + (\not{p}' + M)(\not{p} + M)\gamma^\lambda$$

$$\times (p + p')^\mu) + \frac{(\kappa F_2)^2}{(2M)^2}\text{Tr}((\not{p}' + M)(\not{p} + M))(p + p')^\mu(p + p')^\lambda;$$

The first one:

$$(F_1 + \kappa F_2)^2 Tr((\not{p}' + M)\gamma^\mu(\not{p} + M)\gamma^\lambda)$$

$$= (F_1 + \kappa F_2)^2 4(p'^\mu p^\lambda + p'^\lambda p^\mu + (M^2 - p' \cdot p)g^{\mu\lambda}).$$

The middle one:

$$(F_1 + \kappa F_2)\frac{\kappa F_2}{2M} Tr((\not{p}' + M)\gamma^\mu(\not{p} + M)(p + p')^\lambda + (\not{p}' + M)$$

$$\times (\not{p} + M)\gamma^\lambda(p + p')^\mu) = (F_1 + \kappa F_2)4\kappa F_2(p + p')^\mu(p + p')^\lambda.$$

The last one:

$$\frac{(\kappa F_2)^2}{(2M)^2}Tr((\not{p}' + M)(\not{p} + M)(p + p')^\mu(p + p')^\lambda)$$

$$= \frac{(\kappa F_2)^2}{M^2}((p' \cdot p) + M^2)(p + p')^\mu(p + p')^\lambda.$$

The middle and last can be combined to give the final form:

$$M^{\mu\lambda} = (F_1 + \kappa F_2)^2 4(p'^\mu p^\lambda + p'^\lambda p^\mu + (M^2 - p' \cdot p)g^{\mu\lambda})$$

$$- (4(F_1 + \kappa F_2)\kappa F_2 - \frac{(\kappa F_2)^2}{M^2}(p' \cdot p + M^2))(p + p')^\mu(p + p')^\lambda;$$

in agreement with Equation (5.25). The contraction with the electron tensor to give Equation (5.27) is straightforward.

Problem 5.5. Check the Rosenbluth formula for the $(\kappa F_2)^2$ term. The part of $M^{\mu\lambda}$ of interest is obtained by setting $F_1 = 0$. Then

$$M^{\mu\lambda} = 4(\kappa F_2)^2 \left\{ p'^\mu p^\lambda + p'^\lambda p^\mu + (M^2 - p \cdot p')g^{\mu\lambda} \right.$$
$$\left. + \left(\frac{p \cdot p'}{4M^2} - \frac{3}{4} \right)(p+p')^\mu (p+p')^\lambda \right\}. \qquad (5.103)$$

Contracting with the electron tensor and dropping the factor $(\kappa F_2)^2$ gives

$$E_{\mu\lambda}M^{\mu\lambda} = 32 \left\{ k \cdot p' k' \cdot p + k \cdot p k' \cdot p' - M^2 k \cdot k' \right.$$
$$+ \left(\frac{p \cdot p'}{4M^2} - \frac{3}{4} \right)(k \cdot (p+p')k' \cdot (p+p'))$$
$$\left. - k \cdot k'(p+p')^2/2) \right\}. \qquad (5.104)$$

According to Equation (5.29), this expression should equal

$$64M^2 E E' \left(-\frac{q^2}{4M^2}\cos^2(\theta/2) - \frac{q^2}{2M^2}\sin^2(\theta/2) \right)$$
$$= 4q^2(q^2 - 4EE'). \qquad (5.105)$$

This exercise is to show that this is indeed true. One more intermediate step to check your algebra, and you should be able to take it from here:

$$E_{\mu\lambda}M^{\mu\lambda} = 32 \left\{ \frac{M^2(E-E')^2}{2} - \frac{q^2(E+E')^2}{8} + \frac{q^6}{32M^2} \right\}. \qquad (5.106)$$

There are many equivalent ways to express the invariant forms obtained from the trace theorems. We have shown that Rosenbluth's compact formula is one of them. But we have to admire Rosenbluth's derivation in the first place.

Solution 5.5. First confirm that Equation (5.106) is correct, then obtain it from Equation (5.104). Our first task is to show:

$$8\left(\frac{M^2(E-E')^2}{2} - \frac{q^2(E+E')^2}{8} + \frac{q^6}{32M^2}\right) = q^4 - 4q^2 EE'.$$

All you need to show that this formula is indeed true is $-q^2/(2M) = E - E'$. The first term $4M^2(E-E')^2 = q^4$, leaving the unlikely looking relation:

$$-q^2(E+E')^2 + \frac{q^6}{4M^2} = -4q^2 EE'.$$

However, an overall factor of q^2 cancels, and what is left over becomes $-(E-E')^2 + (E-E')^2 = 0$. So Equation (5.106) is correct, all we have to do is obtain it from (5.104). Use the following formulas:

$$k'\cdot p = k\cdot p' = ME'; \ k\cdot p = k'\cdot p' = ME; \ k\cdot k' = -\frac{q^2}{2};$$

$$p\cdot p' = -\frac{q^2}{2} + M^2; \ (p+p')^2 = 4M^4 - q^2.$$

The factor of 32 out in front cancels, so we drop it. Then (5.104) becomes:

$$M^2(E^2+E'^2) + \frac{q^2 M^2}{2} - \left(\frac{q^2}{8M^2} + \frac{1}{2}\right)$$

$$\times\left(M^2(E+E')^2 + M^2 q^2 - \frac{q^4}{4}\right)$$

$$= \frac{M^2(E-E')^2}{2} - \frac{q^2(E+E')^2}{8} + \frac{q^2}{32M^2}.$$

Problem 5.6. Show that Equation (5.43) satisfies Equation (5.42).

Solution 5.6. This one is easy. Show that Equation (5.43) satisfies (5.42). Equation (5.43) reads:

$$W^{\mu\lambda} = W_1\left(-g^{\mu\lambda} + \frac{q^\mu q^\lambda}{q^2}\right) + \frac{W_2}{M^2}\left(p^\mu - \frac{p\cdot q}{q^2}q^\mu\right)\left(p^\lambda - \frac{p\cdot q}{q^2}q^\lambda\right);$$

and Equation (5.42) is the requirement:

$$q_\mu W^{\mu\lambda} = q_\lambda W^{\mu\lambda} = 0.$$

So

$$q_\mu W^{\mu\lambda} = W_1(-q^\lambda + q^\lambda) + \frac{W_2}{M^2}(p \cdot q - p \cdot q)\left(p^\lambda - \frac{p \cdot q}{q^2}q^\lambda\right) = 0;$$

and $q_\lambda W^{\mu\lambda} = 0$ by a similar argument.

Problem 5.7. Verity Equation (5.44) by contracting $W^{\mu\lambda}$ with the electron tensor, and then use the laboratory frame kinematics to obtain Equation (5.45).

Solution 5.7. Verify Equation (5.44) by contracting $W^{\mu\lambda}$ with Equation (5.26) (not (5.25) — typo) $E_{\mu\lambda} = 4(k'_\mu k_\lambda + k'_\lambda k_\mu - (k' \cdot k)g_{\mu\lambda})$. The W_1 term:

$$32\pi MW_1\left(-g^{\mu\lambda} + \frac{q^\mu q^\lambda}{q^2}\right)(k'_\mu k_\lambda + k'_\lambda k_\mu - (k' \cdot k)g_{\mu\lambda})$$

$$= 32\pi MW_1(k \cdot k')\left(1 - \frac{2k' \cdot k}{q^2}\right) = 64\pi MW_1(k \cdot k').$$

The W_2 term is longer:

$$32\pi\frac{W_2}{M}\left(p^\mu - \frac{p \cdot q}{q^2}q^\mu\right)\left(p^\lambda - \frac{p \cdot q}{q^2}q^\lambda\right)(k_\mu k'_\lambda + k_\lambda k'_\mu - (k \cdot k')g_{\mu\lambda})$$

$$= 32\pi\frac{W_2}{M}\left(p^\mu - \frac{p \cdot q}{q^2}q^\mu\right) \times (k_\mu(k' \cdot p) + k'_\mu(k \cdot p) - p_\mu(k' \cdot k)$$

$$-\frac{q \cdot p}{q^2}(k_\mu(k' \cdot q) + k'_\mu(k \cdot q) - (k' \cdot k)q_\mu)).$$

One more contraction gives Equation (5.44).

$$8\pi ME_{\mu\lambda}W^{\mu\lambda} = 64\pi MW_1(k \cdot k')$$

$$+\frac{32\pi W_2}{M}\left\{2(k \cdot p)(k' \cdot p) - (k \cdot k')p^2\right.$$

$$-\frac{2p \cdot q}{q^2}((k \cdot q)(k' \cdot p) + (k \cdot p)(k' \cdot q) - (k \cdot k')(q \cdot p))$$

$$\left.+\frac{(p \cdot q)^2}{q^4}(2(k \cdot q)(k' \cdot q) - (k \cdot k')q^2)\right\}.$$

To obtain Equation (5.45), the W_1 term is easy: $k \cdot k' = 2EE'\sin^2(\theta/2)$. The W_2 term requires more work. For the target proton at rest:

$$k \cdot p = ME; \ k' \cdot p = ME'; \ p^2 = M^2; \ k \cdot k' = -q^2/2$$
$$= 2EE' \sin^2(\theta/2); q = k - k'; \ q \cdot k = -k \cdot k'; \ q \cdot k'$$
$$= k \cdot k'; \ p \cdot q = M(E - E').$$

With these substitutions the $(p \cdot q/q^2)$ and $(p \cdot q)^2/q^4$ terms both vanish, leaving only:

$$2M^2EE' - M^2(k \cdot k') = 2M^2EE'(1 - \sin^2(\theta/2)) = 2M^2EE' \cos^2(\theta/2).$$

Problem 5.8. Verify the virtual photon polarization sum rule Equation (5.58), and that the $q^\mu q^\nu$ term does not couple to the vector current $\bar{u}(k')\gamma^\lambda u(k)$.

Solution 5.8. Polarization vectors for the exchanged intermediate photon are defined by Equation (5.56). The z axis is chosen along the four momentum q.

$$q^\mu = (\nu, 0, 0, |\vec{k} - \vec{k}'|);$$

and the polarization vectors are

$$\epsilon^\mu_{\lambda = \pm 1} = \mp \frac{1}{\sqrt{2}}(0, 1, \pm i, 0); \ \text{and} \ \epsilon^\mu_{\lambda = 0} = \frac{1}{\sqrt{-q^2}}(|\vec{k} - \vec{k}'|, 0, 0, \nu).$$

These vectors obey the following relations:

$$\epsilon^\mu_\lambda q_\mu = 0; \ \epsilon^{\mu*}_\lambda \epsilon_{\mu, \lambda'} = 0, \lambda \neq \lambda'; \ \epsilon^{\mu*}_\lambda \epsilon_{\mu\lambda} = -1; \lambda = \pm 1; \ \text{and}$$
$$\epsilon^\mu_\lambda \epsilon_{\mu\lambda} = +1; \lambda = 0.$$

The transverse polarization states $\lambda = \pm 1$ are space-like, while the longitudinal state $\lambda = 0$, that does not exist for a real photon, is time-like. This explains the sign change in Equation (5.58).

$$\sum_\lambda (-1)^{\lambda+1} \epsilon^{\mu*}_\lambda \epsilon^\nu_\lambda = -g^{\mu\nu} + \frac{q^\mu q^\nu}{q^2};$$

It is clear from the properties of the four vectors that this equation is correct. The virtual photon has no rest frame, since its velocity $v^2 = (E^2 + E'^2 - 2EE' \cos\theta)/(E^2 + E'^2 - 2EE') \geq 1$. The $q^\mu q^\nu$ term does not couple to the vector current because of the Dirac equation for $E \gg m_e$:

$$q^\mu q^\nu \bar{u}(k') \gamma_\mu u(k) = q^\nu \bar{u}(k')(\not{k} - \not{k}')u(k) = 0.$$

Problem 5.9. Verify the decomposition of the inelastic electron–proton cross-section in terms of the cross-sections for transverse and longitudinal photons, Equations (5.65) and (5.66).

Solution 5.9. Equations (5.65) and (5.66) are easily reconciled with Equation (5.48).

$$\frac{d\sigma}{dE'd\Omega} = \frac{4\alpha^2 E'^2}{q^4} \left\{ 2W_1 \sin^2(\theta/2) + W_2 \cos^2(\theta/2) \right\}$$

$$= \frac{4\pi^2 \alpha}{K} \left\{ (A - B)W_1 + BW_2 \left(1 - \frac{\nu^2}{q^2} \right) \right\};$$

where $\sigma_T = (4\pi^2 \alpha)/K \times A$, and $\sigma_L = (4\pi^2 \alpha)/K \times B$. A little algebra gives:

$$A = \frac{K\alpha E'}{2\pi^2 E|q^2|} \frac{|q^2|\cos^2(\theta/2) + 2\sin^2(\theta/2)(\nu^2 + |q^2|)}{2\sin^2(\theta/2)(\nu^2 + |q^2|)};$$

and

$$B = \frac{K\alpha E'}{2\pi^2 E|q^2|} \frac{|q^2|\cos^2(\theta/2)}{2\sin^2(\theta/2)(\nu^2 + |q^2|)}.$$

The W_1 term:

$$\frac{4\pi^2 \alpha}{K}(A - B) = \frac{8\alpha^2 E'^2 \sin^2(\theta/2)}{q^4}; \text{ or } \frac{2E'}{E|q^2|} = \frac{8E'^2 \sin^2(\theta/2)}{q^4}.$$

This is true because $|q^2| = 4EE' \sin^2(\theta/2)$. The W_2 term is even easier, remembering that $(1 - \nu^2/q^2) = (\nu^2 + |q^2|)/|q^2|$.

Problem 5.10. Obtain Equation (5.80) from Equation (5.79).

Solution 5.10. Equations (5.79) and (5.80) are the same in the limit that all masses are set to zero. However, you have to convert to the invariant variables s, t, u before setting the proton mass to zero.

$$\frac{d\sigma}{dtdu} = \frac{2\pi\alpha^2}{4E^2\sin^4(\theta/2)}\frac{1}{4MEE'}$$

$$\times \left\{\frac{2}{M}F_1(x)\sin^2(\theta/2) + \frac{1}{\nu}F_2(x)\cos^2(\theta/2)\right\};$$

and in the limit $M \to 0$:

$$\frac{d\sigma}{dtdu} = \frac{4\pi\alpha^2}{s^2t^2}\left\{(s+u)xF_1(x) - \frac{us}{s+u}F_2(x)\right\}.$$

We have brought the factor $1/(s+u)$ inside the brackets of Equation (5.80). The denominator outside the brackets is:

$$4E^2\sin^4(\theta/2) \times 4MEE' = 16E^2E'^2\sin^4(\theta/2)\frac{ME}{E'}$$

$$= t^2\frac{ME}{E'} = \frac{t^2s^2}{4MEE'}.$$

Substituting into Equation (5.79) gives:

$$\frac{d\sigma}{dtdu} = \frac{8\pi\alpha^2MEE'}{t^2s^2}\left\{\frac{2}{M}F_1(x)\sin^2(\theta/2) + \frac{1}{\nu}F_2(x)\cos^2(\theta/2)\right\};$$

We are part way there. The F_1 term:

$$\frac{2\sin^2(\theta/2)}{M} = \frac{2}{M}\frac{-t}{4EE'} = \frac{(s+u)x}{2EE'M};$$

leaving us with:

$$\frac{d\sigma}{dtdu} = \frac{4\pi\alpha^2}{t^2s^2}\left\{(s+u)xF_1(x) + \frac{2MEE'\cos^2(\theta/2)}{E - E'}F_2(x)\right\}.$$

The coefficient multiplying $F_2(x)$ is:

$$\frac{2EE'M\cos^2(\theta/2)}{E - E'} = \frac{4M^2EE'\cos^2(\theta/2)}{s+u} = \frac{-us + M^2t}{s+u}.$$

Dropping the M^2t term gives the desired result.

Problem 5.11. Feynman's definition $x = p_i/P$ is given in terms of invariant variables as $x = -t/(s+u)$. So, $F_2(x)$ in Equation (5.85) is directly measurable. Calculate the numerical factor in nanobarns multiplying $F_2(x)$ for 20 GeV electrons incident on a hydrogen target, and scattered electrons at $\theta = 10°$ and $E' = 10$ GeV. Calculate x. Calculate the scattered electron current multiplier $I_s F_2(x) dx dy$ for $100\,\mu$ amp incident beam, and a 50 cm long liquid hydrogen target. (I get $I_s = 36$ picoamps.) Measure the scattered current, know your acceptance widths $dx dy$, and you have $F_2(x)$.

Solution 5.11. The formula of interest is Equation (5.85)

$$\frac{d\sigma}{dx dy} = \frac{2\pi\alpha^2}{t^2} s(1 + (1-y)^2) F_2(x); \tag{1}$$

For the arithmetic, an approximate calculation is sufficient. Take the proton mass to be 1 GeV. Then for $E = 20$ GeV, $s = 41$ GeV^2, and for $E' = 10$ GeV at $\theta = 10^0$ scattering angle $t = -6$ GeV^2 and $u = -19$ GeV^2. So $(1-y)^2 = 0.2$, and $x = 0.27$. The numerical factor for the cross-section is:

$$\frac{2\pi\alpha^2}{t^2} s(1 + (1-y)^2) = 1.7 \times 10^{-31} \text{ cm}^2. \tag{2}$$

The density of liquid hydrogen is $\rho = 0.07$ gm/cm^3, so a 50 cm target has 2×10^{24} protons/cm^2. Then $3.7 \times 10^{-7} \times 100\mu$amp gives a scattered current (to be multiplied by the acceptance $dx dy$) of $I_s = 37$ picoamperes, which is about 2×10^8 electrons/sec.

Problem 5.12. Plot the parton distribution functions for the proton given by Eichten *et al.* [10].

$$xu_v(x) = 1.8x^{0.5}(1 - x^{1.5})^{3.5},$$

$$xd_v(x) = 0.67x^{0.4}(1 - x^{1.5})^{4.5},$$

$$x\bar{u}(x) = x\bar{d}(x) = 0.18(1 - x)^{8.5},$$

$$xs(x) = 0.08(1 - x)^{8.5},$$

$$xg(x) = (2.62 + 9.17x)(1 - x)^{5.9}.$$

The valence quark distributions vanish at $x = 0$, but the others do not. Each of the parton distribution functions, the above formulas divided by x, is singular at $x = 0$. The fits were made for

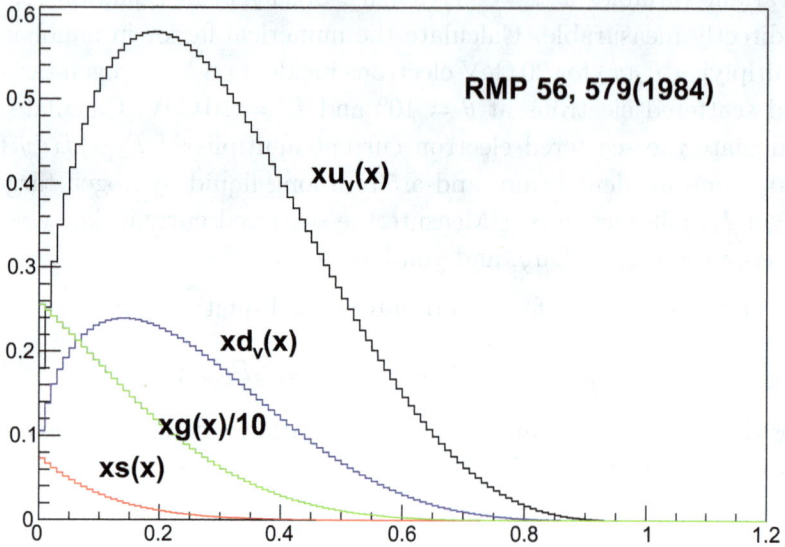

Figure 1: Parton distribution functions for the proton, calculated by Eichten *et al.* The gluon distribution is peaked at low x, and is divided by 10. The $xu(x)$ and $xd(x)$ are valence distributions. The sea distributions $\bar{u}$ and $\bar{d}$ are 2.25 times $s(x)$. So to obtain the total u quark distribution add 2.25 $xs(x)$ to $xu_v(x)$.

$Q^2 = 5$ GeV^2 and $\Lambda = 200$ MeV in the formula for α_s (Equation (5.95)). The Altarelli–Parisi equations were used. Note that the strange quarks are smaller than the up or down quarks in the sea by about a factor or two.

Solution 5.12. The parton distribution functions given in the RMP by Eichten, *et al.* are plotted in Figure 1.

Chapter 6

Problem 6.1. Derive Equation (6.18)

$$\left\langle u^{\dagger(2)}(p)\left|\frac{1-\gamma^5}{2}\right|u^{(2)}(p)\right\rangle = \frac{1}{2}(1+v).$$

Solution 6.1. Prove Equation (6.18):

$$\left\langle u^{\dagger(2)}(p)\left|\frac{1-\gamma^5}{2}\right|u^{(2)}(p)\right\rangle = \frac{E+p}{2E} = \frac{1}{2}(1+v).$$

$$u^{(2)}(p) = \sqrt{E+m}\begin{pmatrix} 0 \\ 1 \\ 0 \\ -p/(E+m) \end{pmatrix};$$

$$u^{\dagger(2)}(p) = \sqrt{E+m}\begin{pmatrix} 0 & 1 & 0 & -p/(E+m) \end{pmatrix};$$

$$\frac{1-\gamma^5}{2} = \frac{1}{2}\begin{pmatrix} 1 & 0 & -1 & 0 \\ 0 & 1 & 0 & -1 \\ -1 & 0 & 1 & 0 \\ 0 & -1 & 0 & 1 \end{pmatrix};$$

$$\left\langle u^{\dagger(2)}(p)\left|\frac{1-\gamma^5}{2}\right|u^{(2)}(p)\right\rangle$$

$$= \frac{1}{2}\left(E+m+2p+\frac{p^2}{E+m}\right)$$

$$= E+p.$$

Divide by the spinor normalization $2E$ to obtain the result.

Problem 6.2. Check Equations (6.23) and (6.24):

$$\text{Tr}(A^\dagger A) = \text{Tr}(B^\dagger B) = 32m_p^2(2E_e E_\nu - p_e \cdot p_\nu)$$
$$\text{Tr}(C^\dagger C) = \text{Tr}(D^\dagger D) = 32m_p^2\lambda^2(2E_e E_\nu + p_e \cdot p_\nu).$$

Solution 6.2. For Equation (6.23):

$$\sum_{\text{spins}} A^\dagger A = \sum_{\text{spins}} (\bar{u}_n\gamma^\lambda u_p)(\bar{v}_\nu\gamma_\lambda u_e)(\bar{u}_p\gamma^\mu u_n)(\bar{u}_e\gamma_\mu v_\nu)$$

$$= \text{Tr}\left\{(\not{p}_n + m_n)\gamma^\lambda(\not{p}_p + m_p)\gamma^\mu\right\} \text{Tr}\left\{\not{p}_\nu\gamma_\lambda(\not{p}_e + m_e)\gamma_\mu\right\}$$

$$= 16m_p^2(2g^{\lambda 0}g^{\mu 0})(p_{\nu\lambda}p_{e\mu} + p_{\nu\mu}p_{e\lambda} - g_{\mu\lambda}(p_e \cdot p_\nu)).$$

We have used $\not{p}_p = \not{p}_n = m_p\gamma^0$. The m_e term does not contribute. Equation (6.23) is the result of the contraction. For $B^\dagger B$:

$$\sum_{\text{spins}} B^\dagger B = \sum_{\text{spins}} (\bar{u}_p\gamma^\mu u_n)(\bar{u}_e\gamma_\mu\gamma^5 v_n)(\bar{u}_n\gamma^\lambda u_p)(\bar{v}_\nu\gamma_\lambda\gamma^5 u_e)$$

$$= \text{Tr}((\not{p}_p + m_p)\gamma^\mu(\not{p}_n + m_n)\gamma^\lambda)$$
$$\times \text{Tr}((\not{p}_e + m_e)\gamma_\mu\gamma^5\not{p}_\nu\gamma_\lambda\gamma^5).$$

The γ^5's cancel, and the remaining expression is the same as $A^\dagger A$. For $C^\dagger C$:

$$\sum_{\text{spins}} C^\dagger C = \lambda^2 \sum_{\text{spins}} (\bar{u}\gamma^\lambda\gamma^5 u_p)(\bar{v}_\nu\gamma_\lambda u_e)(\bar{u}_e\gamma_\mu v_\nu)(\bar{u}_p\gamma^\mu\gamma^5 u_n).$$

The two γ^5's in the neutron-proton current changes a sign on a metric tensor, resulting in a sign change on $(p_e \cdot p_\nu)$.

$$\sum_{\text{spins}} C^\dagger C = \lambda^2 m_p^2 \text{Tr}(\gamma^0\gamma^\lambda\gamma^0\gamma^\mu - \gamma^\lambda\gamma^\mu) \times \text{Tr}(\not{p}_\nu\gamma_\lambda\not{p}_e\gamma_\mu)$$

$$= 16\lambda^2 m_p^2(2g^{0\lambda}g^{0\mu} - g^{00}g^{\mu\lambda} - g^{\mu\lambda})$$
$$\times (p_{\nu\lambda}p_{e\mu} + p_{e\lambda}p_{\nu\mu} - g_{\mu\lambda}(p_e \cdot p_\nu)).$$

Setting $g^{00} = 1$ gives the result:

$$\sum_{\text{spins}} C^\dagger C = 32\lambda^2 m_p^2 (2E_e E_\nu + p_e \cdot p_\nu).$$

$D^\dagger D$ has more γ^5's, but the result is the same. The γ^5's in the lepton tensor go away without harm.

$$\sum_{\text{spins}} D^\dagger D = \lambda^2 \sum_{\text{spins}} (\bar{u}_n \gamma^\mu \gamma^5 u_p)(\bar{u}_p \gamma^\lambda \gamma^5 u_n)(\bar{v}_\nu \gamma_\mu \gamma^5 u_e)(\bar{u}_e \gamma_\lambda \gamma^5 v_\nu)$$

$$= \lambda^2 m_p^2 Tr(\gamma^0 \gamma^\mu \gamma^0 \gamma^\lambda - \gamma^\mu \gamma^\lambda) \times Tr(\not{p}_\nu \gamma_\mu \not{p}_e \gamma_\lambda).$$

So $C^\dagger C$ and $D^\dagger D$ are identical.

Problem 6.3. Obtain Equation (6.28) for ^{14}O decay from Equation (6.27).

Solution 6.3. Equation (6.27) for $^{14}O \rightarrow {}^{14}N^* + e^+ + \nu_e$ reads:

$$d\Gamma_{i \rightarrow f} = \frac{2G_F^2 \cos^2 \theta_c}{(2\pi)^5} \delta(Q - E_e - E_\nu) d^3 p_e d^3 p_\nu (1 + |v_e| \cos \theta_{e\nu}).$$

Note that there is no momentum conservation δ function, because the parent and daughter nuclei are both at rest, and the positron and neutrino have to conserve energy, but not momentum. Integration over the two solid angles eliminates the $\cos \theta_{e\nu}$ term. Then integrating over dE_ν removes the δ function:

$$\frac{d\Gamma}{dE_e} = \frac{G_F^2 \cos^2(\theta_c)}{\pi^3} \sqrt{E^2 - m_e^2}(Q - E_e)^2 E_e.$$

Then introducing the dimensionless variable $x = E_e/Q$ gives Equation (6.28):

$$\frac{d\Gamma}{dx} = \frac{G_F^2 \cos^2(\theta_c)}{\pi^3} Q^5 x (1 - x)^2 \sqrt{x^2 - m_e^2/Q^2}.$$

Problem 6.4. Neutron decay in the beam method. Assume thermal neutrons all with velocity 2000 m/sec and lifetime 879 s. What fraction decays in a 2-m long pipe? For a neutron current of 10^9 neutrons/s through the pipe, and a detection efficiency for the daughter protons of 10%, how many proton counts per sec do you record? Show that the proton recoil kinetic energy is about 750 eV (assumed to be zero in our analysis). The shape of the recoil proton energy distribution depends on the angular correlation parameter $a = (1 - \lambda^2)/(1 + 3\lambda^2)$. To show this, it is necessary to consider the neutron decay as a three-body rather than a two-body final state. We lead the reader through the problem, but you are expected to confirm the calculations. First write the energy dependence of Equation (6.25) as

$$\sum_{\text{spins}} |\langle pe^-\bar{\nu}_e| T |n\rangle|^2 \sim (E_e E_\nu + ap_e E_\nu \cos(\theta_{e\nu})).$$

Now use Equation (2.67) for $2p_e E_\nu \cos(\theta_{e\nu})$:

$$2p_e E_\nu \cos(\theta_{e\nu}) = m_n^2 + m_e^2 - m_p^2 + 2E_e E_\nu - 2m_n(E_e + E_\nu),$$

$$\text{where } E_\nu = m_n - E_e - m_p - T_p = \Delta - E_e - T_p.$$

T_p is the proton kinetic energy, which as we have seen is about 10^{-6} of the proton mass, so it has to be factored out in order not to get lost. Rearranging terms and approximating $m_n^2 - m_p^2 \sim 2m_n\Delta$ gives:

$$\sum_{\text{spins}} |\langle pe^-\bar{\nu}_e| T |n\rangle|^2 \sim 2E_e(\Delta - E_e - T_p) + a(m_e^2 - 2E_e^2$$

$$+ 2E_e(\Delta - T_p) + 2m_n T_p).$$

Now Equation (2.85) shows that the three-body phase space reduces to $dE_e dE_p$, so we have the recoil proton kinetic energy spectrum:

$$\frac{df(T_p)}{dT_p} = \int_{E_e^-}^{E_e^+} (2E_e(\Delta - E_e - T_p)$$

$$+ a(m_e^2 - 2E_e^2 + 2E_e(\Delta - T_p) + 2m_n T_p)dE_e,$$

where the limits $E_e^\pm$ come from solving the quadratic equation that follows from Equation (2.67) for $|\cos\theta_{ep}| = 1$. Show that these limits are (no approximations):

$$E_e^\pm = \frac{A(\Delta - T_p) \pm \sqrt{A^2(\Delta - T_p)^2 - (A^2 + 4m_e^2 T_p(T_p + 2m_p))(\Delta^2 - 2m_n T_p)}}{2(\Delta^2 - 2m_n T_p)},$$

where $A = \Delta^2 + m_e^2 - 2m_n T_p$. This complicated expression is just the solution to a quadratic equation that follows from the boundary curve for a three-body decay with one mass set to zero. Recall the rules from Chapter 2. For a decay $m \to m_1 + m_2 + m_3$, with $m_3 = 0$, particles 1 and 2 have simultaneous maximum energy, when particle 3 is at rest. The maximum energies follow the two-body formula: $E_1 = (m^2 + m_1^2 - m_2^2)/(2m)$. Show that this simultaneous limit is consistent with the above equation. The measurement of $f(T_p)$ is a challenge. A 750 eV proton will not go through tissue paper, so everything has to be evacuated. An accelerating voltage of 20 kV or so can be added to boost the proton energy, and then measure the distribution riding on 20 kV. The asymmetry parameter is small, so the effect on $f(T_p)$ is also modest.

Solution 6.4. Neutron decay by the beam method. For a 2 m long pipe and $v = 2000 \, \mathrm{m/sec}$ each neutron spends 1 millisec in the pipe, giving a decay probability of $10^{-3}/879 = 1.1 \times 10^{-6}$. Call it 10^{-6}. If the neutron flux is 10^9 n/sec, the number of neutrons in the pipe is $10^9 \times 2/2000 = 10^6$. So there is one decay per second. A proton detection efficiency of 10% gives one proton every 10 sec. So to observe 10^6 recoil protons requires about one year of running at 30% efficiency. (one year $= \pi \times 10^7$ sec.) The maximum proton recoil energy occurs when the neutrino is at rest:

$$E_p^{\max} = \frac{m_n^2 + m_p^2 - m_e^2}{2m_n}; \quad T_p = E_p - m_p = \frac{(m_n - m_p)^2 - m_e^2}{2m_n}.$$

Numbers $m_p = 938.27$; $m_n = 939.57$; $m_e = 0.511$; all in MeV give the maximum proton kinetic energy $T_p^{\max} = 750$ eV. This is a very small maximum value. The range in air at STP is less than 1 mm. The measurement of the energy spectrum of recoil protons from neutron decay is a challenge. We are asked to obtain a formula for the recoil proton kinetic energy as a function of the electron-neutrino angular correlation parameter $a = (1 - \lambda^2)/(1 + 3\lambda^2)$. The energy dependence of the square of the T matrix element is:

$$\sum_{\text{spins}} |\langle pe^- \bar{\nu}_e | T | n \rangle|^2 \sim E_e E_\nu + a p_e E_\nu \cos(\theta_{e\nu}).$$

For a three-body final state the angular term is given by kinematics:

$$2 p_e E_\nu \cos(\theta_{e\nu}) = m_n^2 + m_e^2 - m_p^2 + 2 E_e E_\nu - 2 m_n (E_e + E_\nu); \quad \text{where}$$

$$E_\nu = m_n - m_p - E_e - T_p = \Delta - E_e - T_p.$$

Substituting for E_ν and rearranging terms gives:

$$2p_e E_\nu \cos(\theta_{e\nu}) = m_e^2 - \Delta^2 + 2E_e(\Delta - E_e - T_p) + 2m_n T_p.$$

The distribution in recoil proton kinetic energy is:

$$\frac{d\Gamma(T_p)}{dT_p} = \int_{E^-}^{E^+} dE_e(2E_e(\Delta - E_e - T_p) + a(m_e^2 - \Delta^2$$

$$+ 2E_e(\Delta - E_e - T_p) + 2m_n T_p)).$$

For some numbers, $T_p^{\max} = (\Delta^2 - m_e^2)/(2m_n) = 745$ eV, less than 1 part in 10^6 of the proton mass! $\Delta = m_n - m_p = 1.29$ MeV, and E_e is comparable. We make no approximations in evaluating the formulas. This procedure is above suspicion.

$$\int_{E^-}^{E^+} dE(2E(\Delta - E) + a(m_e^2 - \Delta^2 + 2E_e(\Delta - E_e - T_p) + 2m_n T_p))$$

$$= E\left(E\Delta - \frac{2}{3}E^2\right) + aE(m_e^2 - \Delta^2 - \frac{2}{3}E^2 + E(\Delta - T_p)$$

$$+ 2m_n T_p))|_{E^-}^{E^+}.$$

The exact solutions to the quadratic boundary equation in the E_e, E_p plane are given in the text for this problem. The normalized phase space plot that follows from these two solutions is shown in Figure 1. The dependence of the proton kinetic energy distribution on the $\vec{p}_e \cdot \vec{p}_\nu$ angular correlation is shown in Figure 2. The true value of $a = -0.1$, so the effect is even smaller than shown in the figure, and the proton's tiny kinetic energy makes the measurement a challenge. An example of an early measurement of this parameter from decaying neutrons in a vacuum pipe is: C. Stratatowa, R. Dobrozensky, and P. Weinzierl, *Phys Rev* **18**, 3970 (1978). Decay protons traveled down the pipe with the neutron beam, and were energy analyzed in an electrostatic spectrometer.

Problem 6.5. Derive the $V - A$ trace theorems Equations (6.35) and (6.36).

Solution 6.5. Equations (6.35) and (6.36) are adequately derived in the text, and no repetition here is needed. Just go through the steps for yourself.

Electron-proton phase space

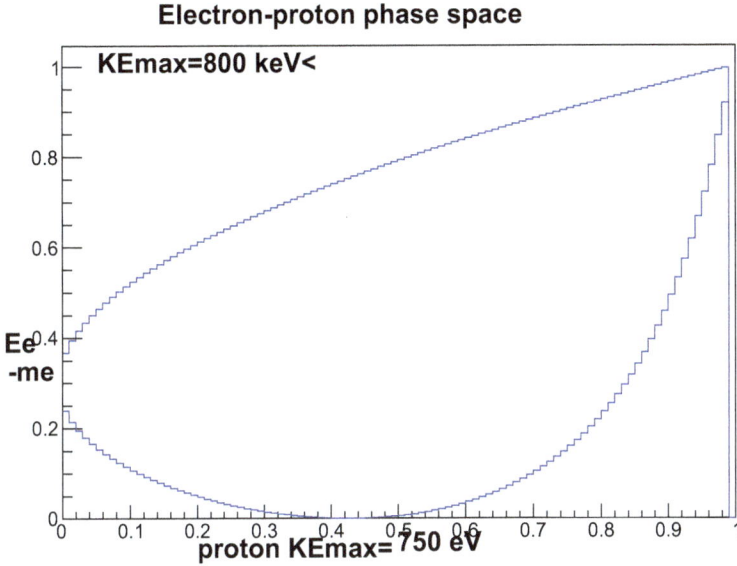

Figure 1: Normalized phase space boundary curve for neutron beta decay. This is the analog of Figure 2 for $K \to \mu\pi\nu$.

Asymmetry of proton KE

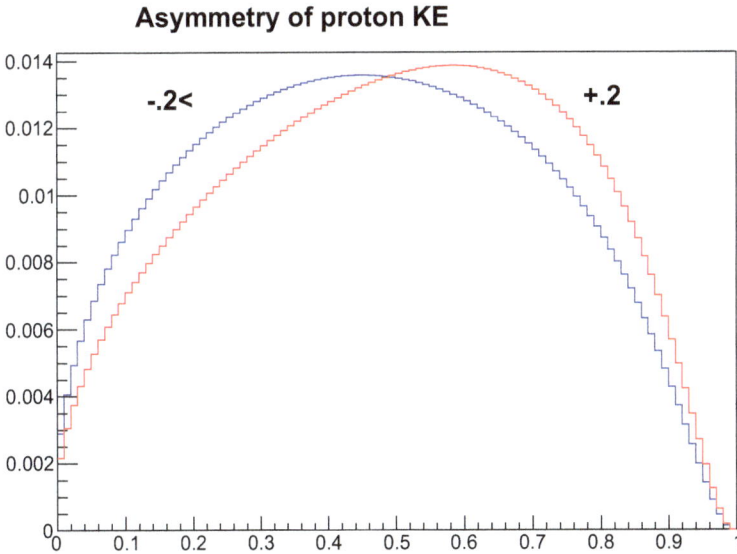

Figure 2: The recoil proton energy spectra for $a = \pm0.2$. The experimental value of a is close to -0.1. The effect is small, but the shift is anticipated. Positive a means that the electron and antineutrino like to come out parallel, and that shifts the recoil protons to higher T_p.

Problem 6.6. Derive the useful identity for reducing a $V-A$ expression that includes a spin vector, Equation (6.49):

$$(1 - \gamma^5)(\not{p}_1 + m)(1 + \lambda_1\gamma^5\not{s}_1)(1 + \gamma^5)$$
$$= (1 - \gamma^5)(\not{p}_1 - \lambda_1 m\not{s}_1 + m)(1 + \gamma^5).$$

Solution 6.6. Equation (6.49) is useful.

$$(1 - \gamma^5)(\not{p}_1 + m)(1 + \lambda_1\gamma^5\not{s}_1)(1 + \gamma^5)$$
$$= (1 - \gamma^5)(\not{p}_1 - \lambda_1 m\not{s}_1 + m)(1 + \gamma^5).$$

The mass term on the right side vanishes. Expand the left side:

$$(1 - \gamma^5)\not{p}_1(1 + \lambda_1\gamma^5\not{s}_1)(1 + \gamma^5) + m(1 - \gamma^5)(1 + \lambda_1\gamma^5\not{s}_1)(1 + \gamma^5)$$
$$= (1 - \gamma^5)\not{p}_1(1 + \gamma^5) - m\lambda_1(1 - \gamma^5)\not{s}_1(1 + \gamma^5)$$
$$+ m(1 - \gamma^5)(1 + \gamma^5).$$

This is equal to the right side.

Problem 6.7. Prove Equation (6.53). Theorem for two-body phase space of massless particles in a three-body final state $p_1 = p_2 + p_3 + p_4$:

$$\int d_2(PS) = \int \delta^4(X - p_2 - p_3)\frac{d^3p_2}{2E_2}\frac{d^3p_3}{2E_3},$$

$$\int d_2(PS)p_2^\alpha p_3^\beta = \frac{\pi}{24}\{g^{\alpha\beta}X^2 + 2X^\alpha X^\beta\},$$

$$X^\alpha = p_1^\alpha - p_4^\alpha = p_2^\alpha + p_3^\alpha.$$

$d_2(PS)$ is Lorentz invariant, and depends only on p_2 and p_3, so the most general form must be:

$$\int d_2(PS)p_2^\alpha p_3^\beta = Ag^{\alpha\beta}X^2 + BX^\alpha X^\beta,$$

$$\text{where } A \text{ and } B \text{ are constants.} \qquad (6.105)$$

Calculate the phase space integrals $\int d_2(PS)p_2 \cdot p_3$, and $\int d_2(PS)(p_2 \cdot X)(p_3 \cdot X)$, giving two equations to solve for A and B. Since the expressions are Lorentz invariant, you may work in the two-neutrino center of mass, where $\vec{p}_2 = -\vec{p}_3$, simplifying the integrals.

Solution 6.7. Proof of Equation (6.53). A two-body phase space for massless particles p_2 and p_3 in the three-body decay $m_1 \rightarrow E_2 + E_3 + E_4$. The phase space integral is defined as:

$$\int d_2(PS) = \int \delta^4(X - p_2 - p_3) \frac{d^3 p_2}{2E_2} \frac{d^3 p_3}{2E_3}.$$

Equation (6.53) reads:

$$\int d_2(PS) p_2^\alpha p_3^\beta = \frac{\pi}{24} \left\{ g^{\alpha\beta} X^2 + 2X^\alpha X^\beta \right\}.$$

The four vector X is:

$$X^\alpha = p_2^\alpha + p_3^\alpha = p_1^\alpha - p_4^\alpha.$$

The derivation is based on the recognition that the left-hand side of Equation (6.53) is Lorentz invariant, and can depend only on X and invariant constants A and B:

$$\int d_2(PS) p_2^\alpha p_3^\beta = A g^{\alpha\beta} X^2 + B X^\alpha X^\beta.$$

Multiply both sides by the metric tensor $g_{\alpha\beta}$:

$$\int \delta^4(X - p_2 - p_3) \frac{d^3 p_2}{2E_2} \frac{d^3 p_3}{2E_3} (p_2 \cdot p_3) = (4A + B)X^2;$$

$$\int \delta^4(X - p_2 - p_3) \frac{d^3 p_2}{2E_2} \frac{d^3 p_3}{2E_3} (E_2 E_3 - \vec{p}_2 \cdot \vec{p}_3)$$

$$= 2\pi \int \delta(X_0 - 2E) E^2 dE = \pi \frac{X^2}{4} = (4A + B)X^2.$$

Here E is the energy of either neutrino in the $\nu\bar{\nu}$ center of mass, where $\vec{X} = 0$. This gives one equation in two unknowns. The second equation can be obtained by multiplying through by $X_\alpha X_\beta$:

$$\int d_2(PS)(p_2 \cdot X)(p_3 \cdot X) = (A + B)X^4;$$

where again in the $\nu\bar{\nu}$ center of mass:

$$\int \delta^4(X - p_2 - p_3) 4E^4 \frac{d^3 p_2}{2E} \frac{d^3 p_3}{2E} = \frac{\pi}{8} X^4.$$

We now have two equations in two unknowns:

$$4A + B = \frac{\pi}{4}; \quad A + B = \frac{\pi}{8}; \quad A = \frac{\pi}{24}; \quad B = \frac{\pi}{12}.$$

The more complicated version of this formula, where the masses m_2 and m_3 are not set to zero, is derived in the Appendix, Equation (A.16).

Problem 6.8. Verify the phase space integral Equation (6.67).

$$\int d^3p_1 \delta \left(m_\pi - E_1 - \sqrt{E_1^2 - m_\mu^2} \right) = \frac{4\pi p^2 E_1}{m_\pi}.$$

Solution 6.8. Equation (6.67) for $\pi \to \mu\nu$ decay reads:

$$\int d^3p_1 \delta(m_\pi - E_1 - \sqrt{E_1^2 - m_\mu^2}) = \frac{4\pi p^2 E_1}{m_\pi};$$

where p is the momentum of either μ or ν in the pion center of mass: $p = (m_\pi^2 - m_\mu^2)/(2m_\pi)$. Convert $d^3p_1 = p_1 E_1 dE_1 d\Omega$, and integration over the solid angle gives 4π. Then the argument of the δ function $f(E_1) = E_1 + \sqrt{E_1^2 - m_\mu^2} - m_\pi$; $dE_1/df = p/m_\pi$. Then $\int \delta(E_1 + \sqrt{E_1^2 - m_\mu^2} - m_\pi)pE_1 dE_1 = p^2 E_1/m_\pi$.

Problem 6.9. Verify the two-pion phase space integral Equation (6.71)

$$\int \frac{(2\pi)^4}{2m_K(2\pi)^6} \delta^4(p_k - p_1 - p_2) \frac{d^3p_1}{2E_1} \frac{d^3p_2}{2E_2} = \frac{1}{16\pi m_K} \frac{p}{E}.$$

Solution 6.9. Phase space integral for $K \to \pi\pi$, Equation (6.71)

$$\int \frac{(2\pi)^4}{2m_K(2\pi)^6} \delta^4(p_k - p_1 - p_2) \frac{d^3p_1}{2E_1} \frac{d^3p_2}{2E_2} = \frac{1}{16\pi m_K} \frac{p}{E}.$$

$$\int \frac{(2\pi)^4}{2m_K(2\pi)^6} \delta^4(p_k - p_1 - p_2) \frac{d^3p_1}{2E_1} \frac{d^3p_2}{2E_2}$$

$$= \frac{1}{8m_K\pi^2} \int \delta(m_K - E_1 - E_2) \frac{d^3p_1}{4E_1 E_2}$$

$$= \frac{1}{8\pi m_K} \int \delta(m_K - 2E) \frac{pdE}{E} = \frac{1}{16\pi m_K} \frac{p}{E};$$

$$E = m_K/2; \quad p = \sqrt{E^2 - m_\pi^2}.$$

Problem 6.10. Adopt Equation (6.68) to the leptonic decay of the $K^+ \to \mu^+ + \nu_\mu$. Replace $\cos\theta_c f_\pi$ with $\sin\theta_c f_K$. Look up the relevant masses, lifetimes and branching ratios, and calculate the ratio f_K/f_π. I got 1.22.

Solution 6.10. Reconfigure Equation (6.68), for pion decay $\pi \to \mu\nu$ to kaon decay $K \to \mu\nu$. Equation (6.68) reads:

$$\Gamma(\pi \to \mu\nu) = \frac{G_F^2 \cos^2\theta_c f_\pi^2}{8\pi} m_\pi m_\mu^2 \frac{(m_\pi^2 - m_\mu^2)^2}{m_\pi^4}.$$

So:

$$\Gamma(K \to \mu\nu) = \frac{G_F^2 \sin^2\theta_c f_K^2}{8\pi} m_K m_\mu^2 \frac{(m_K^2 - m_\mu^2)^2}{m_K^4}.$$

The numbers are $\Gamma(K \to \mu\nu) = 3.35 \times 10^{-14}$ MeV; $G_F^2 \sin^2\theta_c = 6.7 \times 10^{-24}$ MeV^{-4}; $m_K m_\mu^2 = 5.44 \times 10^6$ MeV^3; giving $f_K = 157$ MeV, compared to $f_\pi = 132$ MeV, or a ratio of 1.2.

Problem 6.11. Decays of the τ lepton. The τ, with a mass $m_\tau = 1777$ MeV has many decay modes, but all measurements are consistent with it being a true lepton, obeying the Dirac equation for a spin 1/2 particle. Therefore, despite the heavy mass, it may be treated just like a muon in the decay $\tau^+ \to e^+ \nu_e \bar\nu_\tau$. The τ lifetime is 2.90×10^{-13} s, and the branching fraction to $e^+ \nu_e \bar\nu_\tau = 17.8\%$. Check these numbers with the muon formula Equation (6.46) (I get the prediction to be about 2% low). Another interesting decay mode is $\tau^+ \to \pi^+ \bar\nu_\tau$, with a branching fraction of 10.8%. The decay $\pi^+ \to \tau^+ \nu_\tau$ is energy forbidden, but $\pi^+ \to \mu^+ \nu_\mu$ is not, and the μ and τ should be interchangeable. So, it is tempting to try our analysis of pion decay on this decay mode of the τ. Write Equation (6.60) as follows:

$$\langle \pi^+ \bar\nu_\tau | T | \tau^+ \rangle = \frac{G_F \cos(\theta_c) f_\pi}{\sqrt{2}} (p_\tau^\lambda - p_\nu^\lambda) \bar{v}(p_\tau) \gamma_\lambda (1 - \gamma^5) u(p_\nu).$$

For pion decay $f_\pi \sim m_\pi$, but here the pion 4-momentum is $q \sim m_\tau$, more than a factor of ten larger. Since the q^2 dependence of f_π is unknown, we soldier on. Show that:

$$\sum_{\text{spins}} |\langle \pi^+ \bar{\nu}_\tau | T | \tau^+ \rangle |^2 = \frac{G_F^2 \cos^2(\theta_c) f_\pi^2}{2} 8 m_\tau^3 E_\nu.$$

Then evaluate the two-body final state phase space (remember to divide by two for the τ spin) to obtain the decay rate:

$$\Gamma(\tau^+ \to \pi^+ \bar{\nu}_\tau) = \frac{G_F^2 \cos^2(\theta_c) f_\pi^2}{4\pi} m_\tau E_\nu^2.$$

Compare this number to the experimental rate, assuming $f_\pi = m_\pi$. The numbers are surprisingly close.

Solution 6.11. Decay of the τ lepton. Formula for the muon decay rate:

$$\Gamma(\mu \to e\nu\bar{\nu}) = \frac{G_F^2 m_\mu^5}{192\pi^3}.$$

As a check, $\Gamma(\mu \to e\nu\bar{\nu}) = 3 \times 10^{-16}$ MeV with this formula and $m_\mu = 105.6$ MeV gives $G_F = 1.16 \times 10^{-11}$ MeV^{-2}. That looks OK. Now $\Gamma(\tau \to e\nu\bar{\nu}) = 4 \times 10^{-10}$ MeV $= 1.33 \times 10^6 \times \Gamma(\mu \to e\nu\bar{\nu})$, while $(m_\tau/m_\mu)^5 = 1.38 \times 10^6$. So the prediction of the muon formula with the τ mass is about 2% larger than the measured value. The next part of the problem is to calculate the rate for $\tau \to \pi\nu_\tau$ and compare it to $\pi \to \mu\nu$. Begin with the T matrix element given in the problem:

$$\langle \pi^+ \bar{\nu}_\tau | T | \tau^+ \rangle = \frac{G_F \cos(\theta_c) f_\pi}{\sqrt{2}} (p_\tau^\lambda - p_\nu^\lambda) \bar{v}(p_\tau) \gamma_\lambda (1 - \gamma^5) v(p_\nu).$$

The antineutrino spinor should be $v(p_\nu)$ rather than $u(p_\nu)$. Sorry for the misprint in the statement of Problem 6.11. There is a lot of cancellation in evaluation of the trace.

$$\sum_{\text{spins}} |\langle \pi\nu | T | \tau \rangle |^2 = \frac{G_F^2 \cos^2 \theta_c f_\pi^2}{2} (p_\tau^\lambda - p_\nu^\lambda)(p_\tau^\mu - p_\nu^\mu)$$

$$\times Tr((\not{p}_\tau - m_\tau)\gamma_\lambda(1 - \gamma^5)\not{p}_\nu \gamma_\mu (1 - \gamma^5))$$

$$= \frac{G_F^2 \cos^2 \theta_c f_\pi^2}{2} (p_\tau^\lambda - p_\nu^\lambda)(p_\tau^\mu - p_\nu^\mu) 2(Tr(\not{p}_\tau \gamma_\lambda \not{p}_\nu \gamma_\mu)$$

$$- Tr(\not{p}_\tau \gamma_\lambda \not{p}_\nu \gamma_\mu \gamma^5))$$

$$= \frac{G_F^2 \cos^2 \theta_c f_\pi^2}{2}(p_\tau^\lambda - p_\nu^\lambda)(p_\tau^\mu - p_\nu^\mu)8(g_{\alpha\lambda}g_{\beta\mu}$$

$$+ g_{\alpha\mu}g_{\beta\lambda} - g_{\alpha\beta}g_{\lambda\mu} - i\epsilon_{\alpha\lambda\beta\mu})p_\tau^\alpha p_\nu^\beta$$

$$= \frac{G_F^2 \cos^2 \theta_c f_\pi^2}{2}8m_\tau^3 E_\nu.$$

We have used $p_\tau = (m_\tau, 0, 0, 0)$, and $p_\nu = (E_\nu, 0, 0, E_\nu)$. The imaginary part vanishes, as it must, and there is cancellation in the real part. The two body decay rate formula is:

$$d\Gamma(\tau \to \pi\nu) = \frac{(2\pi)^4}{4m_\tau}\delta(m_\tau - E_\pi - E_\nu)\delta(\vec{p}_\pi + \vec{p}_\nu)|\langle \pi\nu| T |\tau\rangle|^2$$

$$\times \frac{d^3 p_\pi}{(2\pi)^3 2E_\pi}\frac{d^3 p_\nu}{(2\pi)^3 2E_\nu} = \frac{G_F^2 \cos^2 \theta_c f_\pi^2}{16\pi^2}m_\tau^2\delta$$

$$\times (m_\tau - E_\pi - E_\nu)p_\pi dE_\pi d\Omega.$$

$$\Gamma(\tau \to \pi\nu) = \frac{G_F^2 \cos^2 \theta_c f_\pi^2}{4\pi}m_\tau E_\nu^2.$$

We divided by two to average over τ spins, and used $E_\nu = p_\pi$ to obtain a factor E_ν/m_τ from the δ function. Experimental numbers: $\Gamma(\tau \to \pi\nu) = 0.34\times10^{13}\times6.58\times10^{-22}\times0.11 = 2.5\times10^{-10}$ MeV. The first number is the reciprocal of the lifetime, the second is $\hbar$, and the third is the branching fraction to $\pi\nu$. Evaluating the calculated rate with $f_\pi = m_\pi$ and $E_\nu = (m_\tau^2 - m_\pi^2)/(2m_\tau)$ gives $\Gamma(\tau \to \pi\nu) = 2.7 \times 10^{-10}$ MeV, which agrees with the experimental number to better than 10%, so our T matrix element is a pretty good guess. (If you use $f_\pi = 0.94m_\pi$ from the pion lifetime the agreement is even better.)

Problem 6.12. Using Equations (6.75) and (6.76), show that the proton helicity for unpolarized Λ's is $\langle \sigma \cdot \hat{n}\rangle = \alpha$, where $\hat{n} = \cos\theta\hat{z} + \sin\theta\hat{x}$, is the direction of the proton momentum in the Λ rest frame, and $\alpha = 2SP/(S^2 + P^2)$.

Solution 6.12. Equations (6.75) and (6.76) for the decay proton wave function in the Λ rest frame are: For Λ spin ↑

$$\psi(1/2, +1/2) = SY_0^0 |1/2, +1/2\rangle + P(Y_1^0 |1/2, +1/2\rangle /\sqrt{3}$$
$$- \sqrt{2}Y_1^{+1} |1/2, -1/2\rangle /\sqrt{3});$$

and for Λ spin $\downarrow$:

$$\psi(1/2, -1/2) = SY_0^0 |1/2, -1/2\rangle - P(Y_1^0 |1/2, -1/2\rangle /\sqrt{3}$$
$$- \sqrt{2}Y_1^{-1} |1/2, +1/2\rangle /\sqrt{3}).$$

Let the proton momentum vector lie in the (x, z) plane, so $\phi = 0$. Then the proton spin vector along its momentum is:

$$\vec{\sigma} \cdot \hat{n} = \cos\theta\sigma_z + \sin\theta\sigma_x = \begin{pmatrix} \cos\theta & \sin\theta \\ \sin\theta & -\cos\theta \end{pmatrix}$$

Rewrite the proton wave functions for Λ spin up and down:

$$\psi(1/2, +1/2) = \sqrt{1/4\pi} \begin{pmatrix} S + P\cos\theta \\ P\sin\theta \end{pmatrix}; \text{ and } \psi(1/2, -1/2)$$

$$= \sqrt{1/4\pi} \begin{pmatrix} P\sin\theta \\ S - P\cos\theta \end{pmatrix}.$$

$$\int d\Omega(\psi^\dagger\psi(\uparrow) + \psi^\dagger\psi(\downarrow))/2 = S^2 + P^2.$$

Expectation values of the proton spin vector are:

$$\psi^\dagger\vec{\sigma} \cdot \hat{n}\psi(\uparrow) = \frac{1}{4\pi}((P^2 + S^2)\cos\theta + 2PS); \text{ and } \psi^\dagger\vec{\sigma} \cdot \hat{n}\psi(\downarrow)$$

$$= \frac{1}{4\pi}(2PS - (P^2 + S^2)\cos\theta).$$

Therefore, for unpolarized Λ's the average proton helicity is:

$$< \vec{\sigma} \cdot \hat{n} > = \frac{2SP}{S^2 + P^2} = \alpha.$$

Problem 6.13. One can demonstrate from the machinery we have developed that a non-collinear Lorentz transformation on a massless Dirac spinor does not alter the helicity. The relevant formulas are all in Chapter 2. Begin with a massless spinor with momentum along $\hat{z}$ and $\langle \vec{\sigma} \cdot \hat{p}\rangle = +1$. Lorentz transform along $\hat{x}$. Now $p' = (\gamma E, v\gamma E, 0, E)$, and

$$\vec{\sigma} \cdot \hat{p}' = \begin{pmatrix} 1/\gamma & v \\ v & -1/\gamma \end{pmatrix}.$$

The spinor transformation matrix $S = \cosh(\chi/2) + \sinh(\chi/2)\gamma^0\gamma^1$, Equation (2.177) written for a transformation along $\hat{x}$. The rapidity χ is defined by $\tanh\chi = v$. Transforming the spinor gives:

$$Su(p) = \begin{pmatrix} \cosh(\chi/2) & 0 & 0 & \sinh(\chi/2) \\ 0 & \cosh(\chi/2) & \sinh(\chi/2) & 0 \\ 0 & \sinh(\chi/2) & \cosh(\chi/2) & 0 \\ \sinh(\chi/2) & 0 & 0 & \cosh(\chi/2) \end{pmatrix}$$

$$\times \sqrt{E} \begin{pmatrix} 1 \\ 0 \\ 1 \\ 0 \end{pmatrix} = \sqrt{E} \begin{pmatrix} \cosh(\chi/2) \\ \sinh(\chi/2) \\ \cosh(\chi/2) \\ \sinh(\chi/2) \end{pmatrix}.$$

The upper and lower components are the same. Show that $\vec{\sigma} \cdot \hat{p}' Su(p) = Su(p)$, an exercise in hyperbolic functions that proves that the spin follows the momentum.

Solution 6.13. This is an exercise in Lorentz transformations of Dirac spinors. The outline of the problem is clear. All one has to do is verify the following hyperbolic function identities:

$$\frac{\cosh(\chi/2)}{\cosh\chi} + \sinh(\chi/2)\tanh\chi = \cosh(\chi/2); \text{ and } \tanh\chi\cosh(\chi/2)$$

$$-\frac{\sinh(\chi/2)}{\cosh\chi} = \sinh(\chi/2).$$

Problem 6.14. $|\Delta I| = 1/2$ rule and hyperon decays was the subject of Problem 1.8 in Chapter 1. However, we have since learned that the decays of Λ and Σ hyperons involve S and P waves. From Problem 1.8 we had the three Σ decays in terms of two complex amplitudes:

$$A_0 = \frac{\sqrt{2}(A_{3/2} + A_{1/2})}{3}; \; A_+ = \frac{A_{3/2} - 2A_{1/2}}{3}; \; A_- = A_{3/2}.$$

Giving $\sqrt{2}A_0 + A_+ - A_- = 0$.

The subscript refers to the sign of the pion charge. In addition to the α parameters listed in Table 6.3, the γ parameters are known, but were omitted for lack of space. They are: $\gamma_0 = 0.16$; $\gamma_+ = -0.97$;

and $\gamma_- = 0.98$. Attempt to construct a triangle on a two-dimensional plot with the axes labeled S wave and P wave, using the decay rates, the $|\Delta I| = 1/2$ rule formula, and our knowledge of α and γ. Nature has given us more symmetry in these three decays than is required by the $|\Delta I| = 1/2$ rule.

Solution 6.14. From the Particle Data Group:

Σ^+ mass $= 1189.4$ MeV, lifetime $= 0.80 \times 10^{-10}$ sec.

$\Sigma^+ \rightarrow p\pi^0$ branching fraction $= 51.6\%$, $\alpha_0 = -0.982 \pm 0.014$; $\gamma_0 = 0.16$.

$\Sigma^+ \rightarrow n\pi^+$ branching fraction $= 48.3\%$, $\alpha_+ = 0.068$; $\gamma_+ = -0.98$.

Σ^- mass $= 1197$ MeV, lifetime $= 1.48 \times 10^{-10}$ sec.

$\Sigma^- \rightarrow n\pi^-$ branching fraction 99%, $\alpha_- = -0.068 \pm 0.008$; $\gamma_- = +0.98$.

These numbers give the following magnitudes of the amplitudes (square root of the transition rate) in units of $\sqrt{\text{MeV}}$: $|A_0| = 2.0 \times 10^{-6}$; $|A_+| = 2.0 \times 10^{-6}$; $|A_-| = 2.1 \times 10^{-6}$; with $S = -P$ for $|A_0|$, pure P wave for $|A_+|$, and pure S wave for $|A_-|$. See Figure 3.

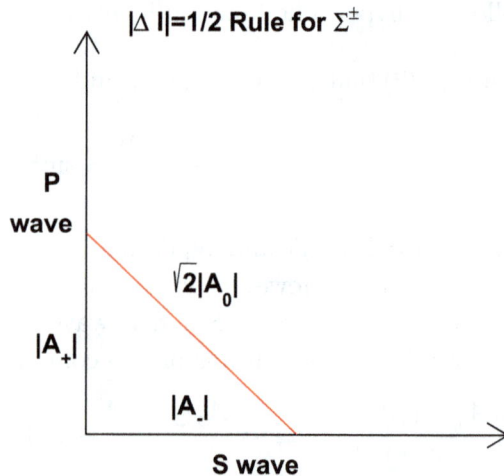

Figure 3: $|\Delta I| = 1/2$ rule for $\Sigma^\pm$ decays. The rule gives the triangle relation $\sqrt{2}A_0 + A_+ - A_- = 0$. Nature has more symmetry than required, since A_- is almost pure "S" wave, A_+ is almost pure "P" wave, A_0 is an equal mixture of the two, and the triangle is right isosceles.

Problem 6.15. Adopt the analysis for $\Lambda \to pe^-\bar{\nu}_e$ to $\Sigma^- \to ne^-\bar{\nu}_e$ and compare with experiment. Use the correct V-spin SU(3) matrix element ($\langle K^0|\,V_+\,|\pi^-\rangle$ in Table 1.4), and $\sin\theta_c$.

Solution 6.15. Hyperon beta decays. The T matrix element for $\Lambda \to pe^-\bar{\nu}_e$ is given by Equation (6.97):

$$\langle pe^-\bar{\nu}_e|\,T\,|\Lambda\rangle = \frac{G_F \sin(\theta_c)}{\sqrt{2}}\sqrt{3/2}(\bar{u}_p\gamma^\mu(1-\gamma^5)u_\Lambda)\times(\bar{u}_e\gamma_\mu(1-\gamma^5)v_{\nu_e}).$$

For $\Sigma^- \to ne^-\bar{\nu}_e$ we write

$$\langle ne^-\bar{\nu}_e|\,T\,|\Sigma^-\rangle = \frac{G_F \sin(\theta_c)}{\sqrt{2}}(\bar{u}_n\gamma^\mu(1-\gamma^5)u_\Sigma)\times(\bar{u}_e\gamma_\mu(1-\gamma^5)v_{\nu_e}).$$

$$(1)$$

The $SU(3)$ matrix element connecting Σ^- to n is given in Table 1.4 as one. Based on the Λ analysis we expect the Σ^- beta decay rate to scale as:

$$\frac{\Gamma(\Sigma^- \to ne^-\bar{\nu}_e)}{\Gamma(\Lambda \to pe^-\bar{\nu}_e)} = \frac{(m_\Sigma - m_n)^5}{(m_\Lambda - m_p)^5} \times \frac{2}{3} \times \frac{1+3\lambda_\Sigma^2}{1+3\lambda_\Lambda^2} = 2.37$$

The measured rate ratio is 2.13, about 10% low. The numbers are all given by the Particle Data Group: $\lambda_\Sigma = 0.34$ and $\lambda_\Lambda = -0.72$.

Problem 6.16. Many baryons containing charm or bottom quarks have been found by experiment. The baryons with charm quarks are listed in the text: $\Lambda_c^+ = udc$, $\Sigma_c^{++} = uuc$; $\Sigma_c^+ = udc$; $\Sigma_c^0 = ddc$; $\Xi_c^+ = usc$; $\Xi_c^0 = dsc$; and $\Omega_c^0 = ssc$, all with spin/parity $(1/2)^+$ and $C = +1$. Note that the Ω_c^0 has spin $1/2$, different from the one with three strange quarks, $\Omega^-(sss)$, that has spin $3/2$. Make a list with quark content of the bottom quark baryons, using the same notation, adopted from the strange particles. Note that the bottom quark has charge $-1/3$, so that the charges of the bottom baryons match those of the strange baryons. The bottom quark has quantum number $B = -1$. Charm decays to strange, which is allowed, but $b \to c$ and $s \to u$ are both forbidden, and hence have longer lifetimes. Λ_b, Σ_b, Ξ_b, and Ω_b with spin $1/2$ have all been seen. Pick out a non-leptonic final state for Λ_c, Σ_c^{++}, Ξ_c^+, and Ω_c^0, and for the b quark analogs, and draw a quark flow diagram for each decay.

Solution 6.16. From the Particle Data Group:

Charmed Baryons:

Λ_c^+ (udc) $1/2^+$ mass 2286 MeV, lifetime 2.0×10^{-13} sec. Decay mode $\Lambda_c^+ \to \Lambda \pi^+ \pi^0$.

Σ_c^{++} (uuc) mass 2454 MeV; Σ_c^+ (udc) mass 2453 MeV; Σ_c^0 (ddc) mass 2454 MeV; all decay by strong interaction $\Sigma_c \to \Lambda_c^+ \pi$.

Ξ_c^+ (usc) mass 2468 MeV; lifetime 4.6×10^{-13} sec; $\Xi_c^+ \to \Sigma^+ K^- \pi^+$.

Ξ_c^0 (dsc) mass 2470 MeV; lifetime 1.53×10^{-13} sec; $\Xi_c^0 \to \Xi^- \pi^+$.

Ω_c^0 (ssc) mass 2695 MeV; lifetime 2.7×10^{-13} sec; $\Omega_c^0 \to \Omega^- \pi^+ \pi^0$.

Doubly Charmed Baryons:

Ξ_{cc}^{++} (ucc) mass 3622 MeV; lifetime 2.3×10^{-13} sec; Ξ_{cc}^+ (dcc) and Ω_{cc}^+ (scc) no data.

Weak decay lifetimes for charmed baryons are in the 10^{-13} sec range because the $c \to s$ transition is Cabibbo favored.

Bottom Baryons:

Λ_b (udb) $1/2^+$ mass 5619.6 ± 0.17 MeV, lifetime 1.47×10^{-12} sec. Main decay mode $\Lambda_b \to \Lambda_c^+ l^- \bar{\nu}_l + \text{anything} = 10\%$.

Σ_b triplet (uub), (udb), (ddb) $1/2^+$; $m_{\Sigma_b^+} = 5810.56 \pm 0.25$ MeV; $m_{\Sigma_b^-} = 5815.64 \pm 0.27$ MeV. Decay $\Sigma_b \to \Lambda_b \pi$ by strong interaction.

Ξ_b^- (dsb) mass 5797.0 MeV lifetime $1.57 \pm 0.04 \times 10^{-12}$ sec.

Ξ_b^0 (usb) mass 5791.9 MeV lifetime $1.48 \pm 0.03 \times 10^{-12}$ sec.

Quark Flow for $\Sigma_c^{++} \to \Lambda_c^+ \pi^+$

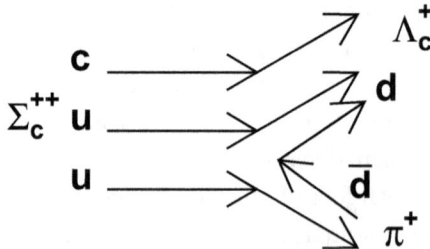

Figure 4:   Quark flow diagram for the strong decay $\Sigma_c^{++} \to \Lambda_c^+ \pi^+$.

Quark Flow for $\Lambda_b^0 \rightarrow \Lambda_c^+ e^- \bar{\nu}_e$

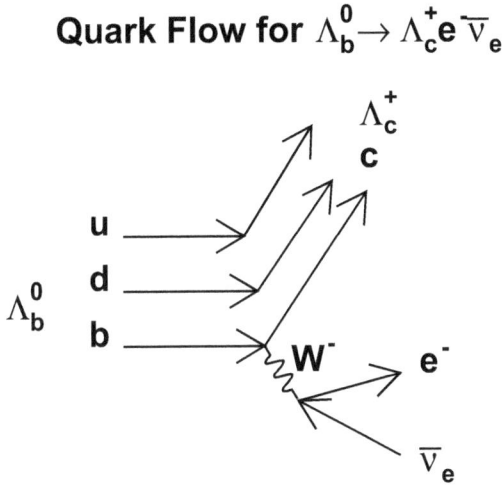

Figure 5: Quark flow diagram for the weak decay $\Lambda_b^0 \rightarrow \Lambda_c^+ e^- \bar{\nu}_e$.

Ω_b^- (ssb) mass 6046.1 ± 1.7 MeV; lifetime $1.64 \pm 0.18 \times 10^{-12}$ sec. Except for the Σ_b's, the lifetimes are all characteristic of the $b \rightarrow c$ lifetime of about 10^{-12} sec, about a factor of 5 longer than $c \rightarrow s$.

There are several quark flow diagrams in the text. Figures 6.6 to 6.10, and 9.16 are examples. Here we will draw two for the heavy quark baryons. Figure 4 shows the decay $\Sigma_c^{++} \rightarrow \Lambda_c^+ \pi^+$ by strong interaction, while Figure 5 shows the weak decay $\Lambda_b^0 \rightarrow \Lambda_c^+ e^- \bar{\nu}_e$.

Chapter 7

Sign errors in the text have been corrected in this statement of the problem.

Problem 7.1. Polarized Møller scattering. Refer to Section 4.9 and Figure 4.4 for the unpolarized analysis. Both initial spins must be longitudinally polarized, so we need two projection operators. For t channel exchange $p_1 \to p_3$ and $p_2 \to p_4$ we have the T matrix amplitude squared:

$$\sum_{\text{spins}} |A(t)|^2 = \frac{e^4}{4t^2} \text{Tr}\{\not{p}_3 \gamma^\mu (\not{p}_1 + m)(1 + \lambda_1 \gamma^5 \not{s}_1) \gamma^\nu\}$$

$$\times \text{Tr}\{\not{p}_4 \gamma_\mu (\not{p}_2 + m)(1 + \lambda_2 \gamma^5 \not{s}_2) \gamma_\nu\}. \quad (7.144)$$

Evaluate the traces to obtain

$$\sum_{\text{spins}} |A(t)|^2 = \frac{2e^4}{t^2} \{p_1 \cdot p_2 p_3 \cdot p_4 + p_1 \cdot p_4 p_2 \cdot p_3$$

$$+ m^2 \lambda_1 \lambda_2 (s_1 \cdot s_2 p_3 \cdot p_4 - s_1 \cdot p_4 s_2 \cdot p_3)\}. \quad (7.145)$$

In the center of mass with $E \gg m$ we have the four vectors: $p_1 = (E, 0, 0, E) = ms_1$ and $p_2 = (E, 0, 0, -E) = ms_2$. Then obtain the expression in terms of the Mandelstam (s, t, u) variables:

$$\sum_{\text{spins}} |A(t)|^2 = \frac{e^4}{2t^2} \{s^2 + u^2 + \lambda_1 \lambda_2 (s^2 - u^2)\}. \quad (7.146)$$

$\sum |A(u)|^2$, for the crossed diagram, is easy. Just interchange t and u in Equation (7.146). The cross term, which gets a minus sign, is not so easy, so it is left for extra credit for those who feel comfortable by now with doing traces all day long. The answer is $e^4 s^2 (1+\lambda_1\lambda_2)/tu$, with a plus sign. Write down the cross-section apart from phase space as a function of λ_1 and λ_2, each of which can be ± 1, for four possibilities. Add them all up and you get the unpolarized cross-section (multiply the spin independent part by four). Show that the formula

$$\sum_{\text{spins}} |A(t) - A(u)|^2 = \frac{e^4}{2} \left\{ \frac{s^2 + u^2 + \lambda_1\lambda_2(s^2 - u^2)}{t^2} \right.$$

$$\left. + \frac{s^2 + t^2 + \lambda_1\lambda_2(s^2 - t^2)}{u^2} + \frac{2s^2}{tu}(1 + \lambda_1\lambda_2) \right\}.$$

$$(7.147)$$

becomes for $\theta = 90°$ scattering angle in the center of mass:

$$\frac{\sigma(m_s = \pm 1)}{\sigma(m_s = 0)} = \frac{1}{8}, \qquad (7.148)$$

where the quantization axis is along the incident beam. This demonstrates the large spin dependent effect for Møller scattering. It has been used to measure the longitudinal polarization of high energy electron beams. See footnote a.

Solution 7.1. Start with Equation (7.144):

$$\sum_{\text{spins}} |A(t)|^2 = \frac{e^4}{4t^2} Tr \left\{ \not{p}_3 \gamma^\mu (\not{p}_1 + m)(1 + \lambda_1 \gamma^5 \not{s}_1) \gamma^\nu \right\}$$

$$\times Tr \left\{ \not{p}_4 \gamma_\mu (\not{p}_2 + m)(1 + \lambda_2 \gamma^5 \not{s}_2) \gamma_\nu \right\}.$$

The factor of 4 in the denominator comes from the spin projection operators. Expand the products:

$$\sum_{\text{spins}} |A(t)|^2 = \frac{e^4}{4t^2} Tr \left\{ \not{p}_3 \gamma^\mu \not{p}_1 \gamma^\nu + m\not{p}_3 \gamma^\mu \gamma^\nu \right.$$

$$\left. + \lambda_1 (\not{p}_3 \gamma^\mu \not{p}_1 \gamma^5 \not{s}_1 \gamma^\nu + m\not{p}_3 \gamma^\mu \gamma^5 \not{s}_1 \gamma^\nu) \right\}$$

---

[a] H. Band *et al.*, *Nucl. Instr. Meth.* A **400**, 24 (1997).

$$\times Tr\left\{\not{p}_4\gamma_\mu\not{p}_2\gamma_\nu + m\not{p}_4\gamma_\mu\gamma_\nu\right.$$

$$\left.+\lambda_2(\not{p}_4\gamma_\mu\not{p}_2\gamma^5\not{s}_2\gamma_\nu + m\not{p}_4\gamma_\mu\gamma^5\not{s}_2\gamma_\nu)\right\}.$$

The traces of the second and third terms in each bracket vanish. The other traces are canonical:

$$\sum_{\text{spins}}|A(t)|^2 = \frac{4e^4}{t^2}\left\{(p_3^\mu p_1^\nu + p_3^\nu p_1^\mu - g^{\mu\nu}p_1\cdot p_3 + i\lambda_1 m\epsilon^{\rho\mu\sigma\nu}p_{3\rho}s_{1\sigma})\right.$$

$$\left.\times(p_{4\mu}p_{2\nu} + p_{4\nu}p_{2\mu} - g_{\mu\nu}p_2\cdot p_4 + i\lambda_2 m\epsilon_{\alpha\mu\beta\nu}p_4^\alpha s_2^\beta)\right\}.$$

The imaginary terms vanish because the four momenta are symmetric under interchange of μ and ν, but ϵ is antisymmetric. So the result is the product of the real parts plus the product of the imaginary parts. The first part is straightforward:

$$(p_3^\mu p_1^\nu + p_3^\nu p_1^\mu - g^{\mu\nu}p_1\cdot p_3)(p_{4\mu}p_{2\nu} + p_{4\nu}p_{2\mu} - g_{\mu\nu}p_2\cdot p_4)$$

$$= 2(p_1\cdot p_2)(p_3\cdot p_4) + 2(p_1\cdot p_4)(p_2\cdot p_3);$$

and the spin terms:

$$-\lambda_1\lambda_2 m^2\epsilon^{\rho\mu\sigma\nu}\epsilon_{\alpha\mu\beta\nu}p_{3\rho}s_{1\sigma}p_4^\alpha s_2^\beta = -\lambda_1\lambda_2 m^2\epsilon^{\mu\nu\sigma\rho}\epsilon_{\mu\nu\beta\alpha}p_{3\rho}s_{1\sigma}p_4^\alpha s_2^\beta$$

$$= 2\lambda_1\lambda_2 m^2(\delta_\beta^\sigma\delta_\alpha^\rho - \delta_\alpha^\sigma\delta_\beta^\rho)p_{3\rho}s_{1\sigma}p_4^\alpha s_2^\beta;$$

where we have used Equation (2.30) to reduce the sum of the product of two ϵ tensors. Assembling the expression gives Equation (7.145) apart from a factor of $1/4$ from the spin average.

$$\sum_{\text{spins}}|A(t)|^2 = \frac{8e^4}{t^2}\left\{(p_1\cdot p_2)(p_3\cdot p_4) + (p_1\cdot p_4)(p_2\cdot p_3)\right.$$

$$\left.+ m^2\lambda_1\lambda_2((p_3\cdot p_4)(s_1\cdot s_2) - (s_1\cdot p_4)(s_2\cdot p_3))\right\}.$$

The Mandelstam variables are: $s = (p_1+p_2)^2 = (p_3+p_4)^2$, $t = (p_1-p_3)^2 = (p_2-p_4)^2$, and $u = (p_1-p_4)^2 = (p_2-p_3)^2$. In the center of mass, if $E \gg m$,

$$s = 2p_1 \cdot p_2 = 4E^2; \quad t = -2p_1 \cdot p_3 = -\frac{s}{2}(1 - \cos\theta); \quad \text{and } u = -2p_1 \cdot p_4$$

$$= -\frac{s}{2}(1 + \cos\theta).$$

Substitution in the above equation gives:

$$\sum_{\text{spins}} |A(t)|^2 = \frac{8e^4}{t^2}\left(\frac{s^2}{4} + \frac{u^2}{4} + m^2\lambda_1\lambda_2\left(\frac{s^2}{4m^2} - \frac{u^2}{4m^2}\right)\right)$$

$$= \frac{2e^4}{t^2}(s^2 + u^2 + \lambda_1\lambda_2(s^2 - u^2))$$

The four vectors are:

$$p_1 = (E, 0, 0, E); \quad s_1 = (E/m, 0, 0, E/m) = p_1/m;$$

$$p_2 = (E, 0, 0, -E); \quad s_2 = (E/m, 0, 0, -E/m) = p_2/m.$$

With this choice for the spin four vectors $\lambda_1\lambda_2 = +1$ means both beam and target electrons have $\vec{\sigma} \cdot \hat{p} = +1$, or both have $\vec{\sigma} \cdot \hat{p} = -1$. In either case $m_s = 0$, because the particles are moving in opposite directions. For $\lambda_1\lambda_2 = -1$, $m_s = \pm 1$. The spin dependent effect favors $m_s = 0$. The crossed diagram has t and u interchanged:

$$\sum_{\text{spins}} |A(u)|^2 = \frac{2e^4}{u^2}(s^2 + t^2 + \lambda_1\lambda_2(s^2 - t^2)).$$

The interference term is difficult to calculate, because it does not factor, and involves many terms. The cross product trace is:

$$-\frac{e^4}{2tu}Tr((\gamma^\mu\not{p}_1 + m\gamma^\mu)(1 + \lambda_1\gamma^5\not{s}_1)\gamma^\nu\not{p}_4(\gamma_\mu\not{p}_2 + m\gamma_\mu)(1 + \lambda_2\gamma^5\not{s}_2)\gamma_\nu\not{p}_3)$$

Multiplying it all out gives four parts. One is spin-independent; one is multiplied by $\lambda_1\lambda_2$, and two that have vanishing trace are multiplied by λ_1 and λ_2 separately. In order to simplify the calculation, we will not write out the two terms that vanish. Since we will be needing them frequently, we repeat the non-trace theorems from Chapter 4

(Equations (4.39)–(4.42)):

$$\gamma^\mu \gamma_\mu = 4; \quad \gamma^\mu \slashed{a} \gamma_\mu = -2\slashed{a}; \quad \gamma^\mu \slashed{a}\slashed{b}\gamma_\mu = 4a\cdot b; \quad \gamma^\mu \slashed{a}\slashed{b}\slashed{c}\gamma_\mu = -2\slashed{c}\slashed{b}\slashed{a}.$$

First look at the spin-independent term:

$$-\frac{e^4}{2tu}Tr((\gamma^\mu \slashed{p}_1 \gamma^\nu \slashed{p}_4 + m\gamma^\mu \gamma^\nu \slashed{p}_4)(\gamma_\mu \slashed{p}_2 \gamma_\nu \slashed{p}_3 + m\gamma_\mu \gamma_\nu \slashed{p}_3))$$

$$= -\frac{e^4}{2tu}Tr(\gamma^\mu \slashed{p}_1 \gamma^\nu \slashed{p}_4 \gamma_\mu \slashed{p}_2 \gamma_\nu \slashed{p}_3 + m\gamma^\mu \slashed{p}_1 \gamma^\nu \slashed{p}_4 \gamma_\mu \gamma_\nu \slashed{p}_3$$

$$+ m\gamma^\mu \gamma^\nu \slashed{p}_4 \gamma_\mu \slashed{p}_2 \gamma_\nu \slashed{p}_3 + m^2 \gamma^\mu \gamma^\nu \slashed{p}_4 \gamma_\mu \gamma_\nu \slashed{p}_3).$$

Applying the above non-trace theorems reduces the trace to:

$$= -\frac{e^4}{2tu}Tr(-8(p_1 \cdot p_2)\slashed{p}_4\slashed{p}_3 + 4m(\slashed{p}_4\slashed{p}_1\slashed{p}_3 + \slashed{p}_2\slashed{p}_4\slashed{p}_3) + 4m^2\slashed{p}_4\slashed{p}_3)).$$

Traces proportional to m vanish. The other two are now straightforward:

$$= -\frac{e^4}{2tu}(-32(p_1 \cdot p_2)(p_3 \cdot p_4) + 16m^2(p_3 \cdot p_4))$$

Dropping the m^2 term at high energy, we have the spin-independent cross-term in Mandelstam variables:

$$= +\frac{4e^4 s^2}{tu}.$$

Now we have to tackle the spin-dependent term, that is similar. There are γ^5's, but they can be canceled. Terms proportional to $m\lambda_1\lambda_2$ have no trace, so we omit them, and only evaluate the $m^2\lambda_1\lambda_2$ term:

$$-\frac{e^4}{2tu}m^2\lambda_1\lambda_2 Tr(\gamma^\mu \gamma^5 \slashed{s}_1 \gamma^\nu \slashed{p}_4 \gamma_\mu \gamma^5 \slashed{s}_2 \gamma_\nu \slashed{p}_3)$$

$$= -\frac{e^4}{2tu}m^2\lambda_1\lambda_2 Tr(\gamma^\mu \slashed{s}_1 \gamma^\nu \slashed{p}_4 \gamma_\mu \slashed{s}_2 \gamma_\nu \slashed{p}_3);$$

the γ^5's cancel out. Applying the non-trace theorems to the remaining expression gives:

$$-\frac{e^4}{2tu}m^2\lambda_1\lambda_2 Tr(-8(s_1 \cdot s_2)\slashed{p}_4\slashed{p}_3) = \frac{4e^4\lambda_1\lambda_2 s^2}{tu}.$$

Assembling the whole works gives Equation (7.147) as amended above. Equation (7.148) gives the correct ratio. It is the $m_s = 0$ spin state that is larger. The author regrets the error in the main text. At 90^0 in the center of mass $t = u = -s/2$, so calculating the cross-section ratio is easy.

Problem 7.2. Solar neutrinos. This is a subject of fundamental importance. There have been several earthbound detectors studying solar neutrinos, but we will consider in some detail only three: Super-Kamiokande-IV, a large water detector in a mine in Japan, the Homestake neutrino detector of the late Ray Davis in Lead, South Dakota, and the Sudbury Neutrino Observatory in a mine in Canada. Begin with Super-Kamiokande-IV, because it is a large water Cherenkov detector that has enough sensitivity to observe $\nu_e + e^- \to \nu_e + e^-$. Solar neutrinos come from fusion reactions in the core of the sun. There are several different fusion chains, the basic one (that has the highest rate but the lowest energy neutrinos) is $p+p+p+p \to {}^4\text{He} + 2e^+ + 2\nu_e$. Note that these are electron neutrinos, not antineutrinos. Antineutrinos come from nuclear reactors. Solar neutrino energies are low, from about 1 to 15 MeV at the upper end. The flux at the earth can be calculated from the standard solar model, and typical numbers are in the range 10^6 to $10^7 \text{cm}^{-2}\text{s}^{-1}$. The flux is directional. It is coming from the sun, and at night it goes through the earth to reach the detector. The fiducial volume of Super-K is 22,500 metric tons of water. Calculate the number of events per day $\nu_e + e^- \to \nu_e + e^-$ for a flux of $5 \times 10^6 \text{ cm}^{-2}\text{s}^{-1}$ 8 MeV electron neutrinos. Each water molecule has 10 electrons. How many events per day if the electron neutrino oscillated into a muon neutrino on the way from the sun to the earth? The cross-section for $\nu_e e^- \to \nu_e e^-$ is isotropic in the center of mass. Sketch the angular distribution of recoil electrons in the water. This distribution has been used by Super-K to point back to the sun. See footnote b. They report a flux of $2.3 \times 10^6 \text{ cm}^{-2}\text{s}^{-1}$.

Solution 7.2. Solar neutrinos in Super-K. Assume $5 \times 10^6 \text{cm}^{-2}\text{sec}^{-1}$ 8 MeV solar neutrinos from ${}^8B$. (Super K's yield is about half this

---

[b]K. Abe *et al.*, (Super-K-IV Collaboration), *Phys. Rev.* D **94**, 052010 (2016).

number). 22 metric kilotons of water is 2.2×10^{10} gm which would be a sphere of radius 18 meters. The real inner detector is a vertical cylinder 36 m high and 34 m in diameter. To simplify the geometry, I assumed a disk with 18 m radius and 24 m thick, oriented facing the sun. This mimics a sphere. The true geometry on a rotating earth is obviously more complicated, and less efficient. The area of my disk is 10^7cm^2, so the neutrino flux is $\Phi = 5 \times 10^{13} \nu/\text{sec}$. The charged current cross-section given in Equation (7.5): $\sigma(\nu_e e^- \rightarrow \nu_e e^-) = 1.3 \times 10^{-43} \text{cm}^2$. The rate of neutrino interactions would be:

$$w(\nu_e e^- \rightarrow \nu_e e^-) = 5 \times 10^{13} \times 1.3 \times 10^{-43}$$

$$\times \frac{6 \times 10^{23} \times 10 \times 2.4 \times 10^3}{18}$$

$$= 5 \times 10^{-3}/\text{sec}.$$

The factor of ten is the number of electrons per water molecule. This is a total of 432 events/day. Super-K's detected yield in this energy range is 15 events/day. Recoil electrons are detected in the water, with a threshold energy of about 3.5 MeV. The angular distribution in the center of mass is isotropic, so the energy distribution of recoil electrons in the lab is flat, and the mean kinetic energy is 3.9 MeV, so only about 1/2 of 8 MeV neutrinos give a recoil above this threshold, and their measured flux is half the number we used, so that gets an expected yield of about 100 events/day. Other losses include detection efficiency, and the average target area and thickness presented to the sun's neutrino beam.

The Lorentz transformation from the center of mass involves the velocity of the center of mass: $v_c = p_{\text{tot}}/E_{\text{tot}} = 8.0/8.511 = 0.94$. $\gamma_c = 2.93$; $s = m_e^2 + 2m_e E_\nu = 8.44 \text{ MeV}^2$; $E_c = (s + m_e^2)/(2\sqrt{s}) = 1.50 \text{ MeV}$; $p_c = (s - m_e^2)/(2\sqrt{s}) = 1.41 \text{ MeV}$; $p_c/E_c = (s - m_e^2)/(s + m_e^2) = E_\nu/(E_\nu + m_e) = v_c$. So the angle map equation is:

$$\tan \theta = \frac{\sqrt{1 - \cos^2 \theta_c}}{\gamma_c(\cos \theta_c + 1)} = \frac{1}{\gamma_c}\sqrt{\frac{1 - \cos \theta_c}{1 + \cos \theta_c}};$$

$\tan \theta$ is never negative. The electron does not appear backwards in the laboratory, where the initial electron is at rest. This is independent of the neutrino energy. In the backward direction $\cos \theta_c = -1$

8 MeV Solar $\nu + e^- \rightarrow \nu + e^-$

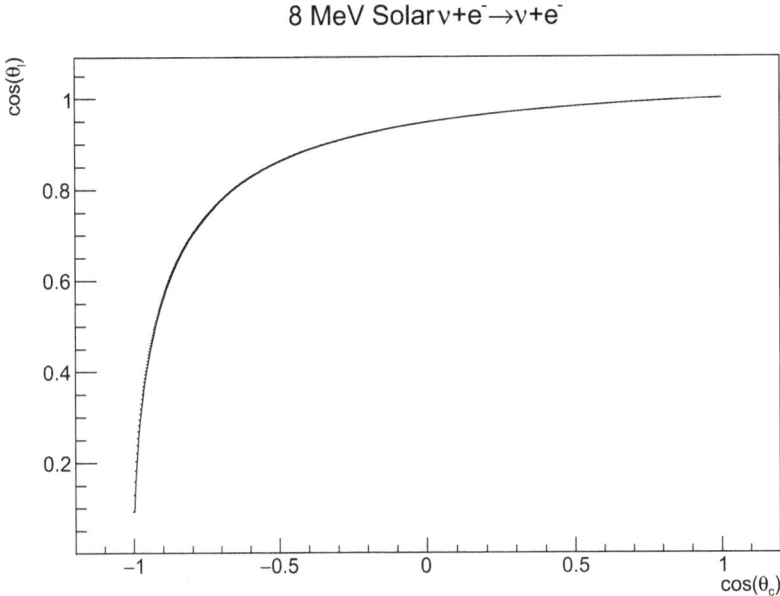

Figure 1: Recoil electron $\cos\theta$ distribution in the laboratory versus $\cos\theta_c$ in the center of mass. $\theta_c = \pi/2$ corresponds to $\theta = 19^0$ relative to the incident neutrino beam, giving a measure of the pointing accuracy of the recoil electrons back to the sun.

the electron transforms to rest in the lab, and is assigned $\theta = \pi/2$. The back scattered neutrino, for which the electron has maximum energy, has an energy of 240 keV. See Figure 1 for the electron angle transformation.

If the electron neutrinos changed flavor on the way to the earth, only the neutral current, which is lepton flavor independent, can be exchanged in the scattering. Equation (7.100) gives the cross-section, about 10% of the charged current, so the observed signal according to Super-K's quoted yield, drops to only a few events per day.

Problem 7.3. Next is Ray Davis. He and a few collaborators operated a tank filled with 378,000 liters of C_2Cl_4, perchloroethylene, density 1.62 gm/cm^3 in a gold mine in South Dakota, nearly one mile underground, for 25 years. This experiment started well before Super-K, but it comes second for us because its cross-section for neutrino detection is not so easy to calculate. It is a nuclear reaction, and you recall that we pointed out at the beginning of our discussion of weak interactions in Chapter 6 that nuclear beta decay lifetimes range from microseconds to millions of years with the same coupling constant G_F. The reaction, proposed for solar neutrino detection by Bruno Pontecorvo, is $\nu_e + {}^{37}\text{Cl} \to {}^{37}\text{Ar} + e^-$. There are two stable isotopes of naturally occurring Chlorine: 25% ^{37}Cl and 75% ^{35}Cl. ^{37}Ar captures a K shell electron to return to ^{37}Cl with a lifetime of 50 days. Argon is a rare gas, so it does not form compounds with the C_2Cl_4, and can be separated out from the tank by using helium gas and other procedures. The threshold neutrino energy for the cross-section is 800 keV, so these neutrinos can be softer than the ones detected by Super-K. The cross-section for Pontecorvo's reaction is $\sigma({}^{37}\text{Cl}(\nu_e, e^-){}^{37}\text{Ar}) = 1.1 \times 10^{-42}\,\text{cm}^2$ for 7 MeV neutrinos from ^{8}B. The flux of these neutrinos is expected to be 5×10^6 cm^{-2}s^{-1}, according to the solar model. First calculate the production rate, and then the equilibrium content of ^{37}Ar in the tank, taking the decay rate into account. Compare this number with the number of ^{37}Cl atoms in the tank. Convert the production rate into Solar Neutrino Units, where one SNU $= 10^{-36}$ captures per target atom per second. Note that the SNU is a transition rate, cross-section times flux. The final result from Homestake is capture rate $= 2.56 \pm 0.16 \pm 0.16$ SNU.[c]

Solution 7.3. Solar neutrinos in the South Dakota gold mine. This is also a cylindrical tank, horizontal in this case. As in Problem 7.2, I assume a disk perpendicular to the sun's rays. Volume is 3.8×10^8 cm^3; radius of the disk is $r = 450$ cm; thickness 600 cm; cross-section for 7 MeV neutrinos $\sigma = 1.1 \times 10^{-42}$cm^2; density is 1.6 gm/cm^3. The molecule C_2Cl_4 molecular weight 168 has four chlorine atoms, but, on average, only one is ^{37}Cl. So the production rate is:

$$w({}^{37}Cl(\nu_e, e^-){}^{37}Ar) = 3 \times 10^{12} \times 1.62 \times 1.1 \times 10^{-42} \times 6$$

$$\times 10^{23} \times 600/168$$

$$= 1.1 \times 10^{-5} \text{ events/sec.}$$

---

[c]B. T. Cleveland *et al.*, *Astrophys. J.* **496**, 505 (1998).

That is about 1 argon atom per day! The number of ^{37}Cl atoms is about 2×10^{30}. The lifetime for electron capture to go back to chlorine is 50 days. So there is an equilibrium amount:

$$\frac{dn}{dt} + \lambda n = w; \; n(t)e^{\lambda t} = \int e^{\lambda t} w dt; \; n(t) = \frac{w}{\lambda}(1 - e^{-\lambda t}); \; n(\infty)$$

$$= \frac{w}{\lambda} = 60.$$

Extracting 60 argon atoms from all of that perchloroethylene deserves a Nobel prize, and Ray Davis got one. The solar neutrino unit, or SNU, is defined as a capture rate: 1 SNU=10^{-36} captures per target atom per second, so this rate is about 5 SNU's. The rate quoted by Davis is half this number. That was the solar neutrino problem.

Problem 7.4. SNO in Canada. Both yields in Problems 7.2 and 7.3 are a factor of two to three below the expectations for the solar neutrino flux. This was known for years as the solar neutrino problem. Possibilities were that the solar model was incorrect, or that the experiments were wrong, or both. It is now generally accepted that theory and experiment are OK, and that the neutrinos themselves are at fault. The electron neutrino changes flavor on its trip to the earth, and muon or tau neutrinos do not scatter as much from electrons, nor do they make ^{37}Ar. The secret to discovering this is the neutral current scattering of neutrinos by nucleons, that does not distinguish between the neutrino flavors, and has a much larger cross-section than neutrino–lepton scattering. The Sudbury Neutrino Observatory was a large sphere 12 m in diameter filled with 1000 metric tons of pure heavy water (much smaller than Super-K, but D_2O rather than H_2O) borrowed from the nuclear power industry in Canada. By detecting neutrons from the breakup of deuterium via $\nu_x + d \rightarrow \nu_x + n + p$, a NC reaction, and comparing that rate to $\nu_e + d \rightarrow e^- + p + p$, a CC reaction, they were able to show that there were neutrinos from the sun that were not electron neutrinos. The deuterium served as both the neutron target and the free neutron detector via the capture $n + {}^2H \rightarrow {}^3H + \gamma$, where $E_\gamma = 6\,\text{MeV}$.

See footnote d. We do not have a satisfactory derivation for either the cross-section for ^{37}Cl $\to$ ^{37}Ar, or for the deuteron break up. Parton model formulas like Equation (7.21) do not hold at 5 MeV neutrino energy. Deuteron breakup requires 2 MeV, and the recoil nucleons have kinetic energy $\sim$1 MeV, hardly the asymptotic high energy limit. The cross-sections predicted by the parton model when extrapolated down to $E_\nu \sim 5$ MeV are too large by about a factor of ten. Accurate derivation of the required breakup formulas is a challenge for theorists. See footnote e for the cross-sections:

$$\text{CC } \sigma(\nu_e d \to e^- pp) \text{ 5 MeV } E_\nu = 0.34 \times 10^{-42} \text{cm}^2;$$

$$\text{8 MeV } E_\nu = 1.44 \times 10^{-42} \text{cm}^2, \qquad (7.149)$$

and

$$\text{NC } \sigma(\nu_e d \to \nu_e np) \text{ 5 MeV } E_\nu = 0.1 \times 10^{-42} \text{cm}^2;$$

$$\text{8 MeV } E_\nu = 0.56 \times 10^{-42} \text{cm}^2. \qquad (7.150)$$

The CC cross-section is only for electron neutrinos, while the NC cross-section is the same for electron, muon or tau neutrinos. Note that the cross-sections do not scale $\sim E_\nu$ because of threshold effects, and that the ratio NC/CC $\sim 1/3$. Take the 8-MeV numbers, and calculate the number of CC and NC events per day for the solar flux given in Problem 7.2, assuming no neutrino flavor change. SNO quotes a detection efficiency for the NC events of 14%, and 576 ± 49 NC events detected for 306 days running. Compare this with your calculated NC yield. The quoted depleted CC flux $1.76 \pm 0.05 \pm 0.09 \times 10^6$ cm^{-2}s^{-1} together with the NC rate that is in agreement with the solar model was clear evidence for neutrino flavor changing. The work outlined in these three problems won a Nobel Prize in physics.

Solution 7.4. The SNO detector and hadronic neutral currents resolved the mystery of where the solar neutrinos went. Leptonic neutral current cross-sections are very small, but for the deuterium

---

[d]Q. R. Ahmad *et al.*, Sudbury neutrino observatory, *Phys. Rev. Lett.* **89**, 011301 (2002).

[e]M. Butler *et al.*, *Phys. Rev. C* **63**, 035501 (2001).

reactions measured by SNO the ratio NC/CC ~ 1/3. For electron neutrinos that oscillated to a different flavor on the way to the earth the flavor independent neutral current cross-section could still be measured, and SNO confirmed that all of the solar neutrinos were there, but 2/3 of them had a different flavor. This detector actually was a sphere, so you do not need to make a simplifying assumption! Remember the old joke about the physicist and the spherical chicken. Here the numbers work out exactly. Mass of D2O $= 10^9$ gm, density $=1.11$ gm/cm^3, volume $= 9 \times 10^8$ cm^3, and area$=1.1 \times 10^6$cm^2, thickness$=820$ cm. So the 8 MeV NC (independent of neutrino flavor) rate is:

$$w(\nu_e + d \to \nu_e np) = 5.5 \times 10^{12} \times .56 \times 10^{-42} \times 1.1 \times 2 \times 6$$

$$\times 10^{23} \times 820/20 = 1.7 \times 10^{-4} \text{ events/sec}.$$

The factor of two is the number of deuterons per molecule. Using the yield and detection efficiency quoted by SNO: 576 events in 306 days and 14% scales to $576/(306 \times 0.14)=13$ events/day, while the above predicted rate is 15 events/day, consistent within statistical errors. Hence all of the solar neutrinos are there, but they are not all still electron neutrinos. The work outlined in these three problems won the Nobel Prize in physics in 2002.

Problem 7.5. Work out the electron mass term for the NC scattering $\nu_\mu e^- \to \nu_\mu e^-$, given in the text after Equation (7.96), which with the extra term is

$$\frac{1}{2} \sum_{\text{spins}} |\langle \nu_\mu e^- | T | \nu_\mu e^- \rangle|^2 = 16 G_F^2 (g_L^2 s^2 + g_R^2 u^2 + 2m^2 g_L g_R t).$$

$$(7.151)$$

The Mandelstam variables keeping the electron mass satisfy $s + t + u = 2m^2$. Show that the dimensionless invariant variable $y = 1 - (p_2 \cdot p_3)/(p_1 \cdot p_2) = -t/(s - m^2)$.

Solution 7.5. We are asked to derive the mass term left out of Equation (7.96), included in Equation (7.151):

$$\frac{1}{2} \sum_{\text{spins}} |\langle \nu_\mu e^- | T | \nu_\mu e^- \rangle|^2 = 16 G_F^2 (g_L^2 s^2 + g_R^2 u^2 + 2m^2 g_L g_R t).$$

Return to Equation (7.93), and look at the cross term including the electron mass; p_1 and p_3 are neutrinos and p_2 and p_4 are electrons, with a mass:

$$A = \bar{u}(p_3)\gamma^{\mu}(1 - \gamma^5)u(p_1)\bar{u}(p_4)\gamma_{\mu}(1 - \gamma^5)u(p_2);$$

$$B = \bar{u}(p_3)\gamma^{\nu}(1 - \gamma^5)u(p_1)\bar{u}(p_4)\gamma_{\nu}(1 + \gamma^5)u(p_2).$$

$$\sum_{\text{spins}} g_L g_R (A^{\dagger} B) = \sum_{\text{spins}} g_L g_R \bar{u}(p_1)\gamma^{\mu}(1 - \gamma^5)u(p_3)\bar{u}(p_3)\gamma^{\nu}$$

$$\times (1 - \gamma^5)u(p_1) \times \bar{u}(p_4)\gamma_{\nu}(1 + \gamma^5)u(p_2)\bar{u}(p_2)$$

$$\times \gamma_{\mu}(1 - \gamma^5)u(p_4).$$

$$= \sum_{\text{spins}} g_L g_R \bar{u}(p_1)\gamma^{\mu}(1 - \gamma^5)\displaystyle{\not}p_3\gamma^{\nu}(1 - \gamma^5)u(p_1)$$

$$\times \bar{u}(p_4)\gamma_{\nu}(1 + \gamma^5)(\displaystyle{\not}p_2 + m)\gamma_{\mu}(1 - \gamma^5)u(p_4).$$

It is clear from this formula that without the mass the cross term vanishes, so we may drop the $\displaystyle{\not}p_2$ part, and $(1 - \gamma^5)^2 = 2(1 - \gamma^5)$, $(1 + \gamma^5)^2 = 2(1 + \gamma^5)$. Thus:

$$\sum_{\text{spins}} (A^{\dagger} B) = 4m g_L g_R \sum_{\text{spins}} \bar{u}(p_1)\gamma^{\mu}(1 - \gamma^5)\displaystyle{\not}p_3\gamma^{\nu}u(p_1)$$

$$\times \bar{u}(p_4)\gamma_{\nu}(1 + \gamma^5)\gamma_{\mu}u(p_4).$$

$$= 4m g_L g_R Tr(\displaystyle{\not}p_1\gamma^{\mu}(1 - \gamma^5)\displaystyle{\not}p_3\gamma^{\nu})Tr((\displaystyle{\not}p_4 + m)$$

$$\times \gamma_{\nu}(1 + \gamma^5)\gamma_{\mu}) = 16m^2 Tr(\displaystyle{\not}p_1\gamma^{\mu}(1 - \gamma^5)\displaystyle{\not}p_3\gamma^{\nu})g_{\mu\nu}.$$

The presence of the metric tensor eliminates the γ^5 trace.

$$\sum_{\text{spins}} (A^{\dagger} B) = 64 g_L g_R m^2 (p_1^{\mu}p_3^{\nu} + p_1^{\nu}p_3^{\mu} - (p_1 \cdot p_3)g^{\mu\nu})g_{\mu\nu}$$

$$= -128 g_L g_R m^2 (p_1 \cdot p_3) = 64 g_L g_R m^2 t.$$

There are two equal cross terms, when divided by two for spin average cancels. For the relative normalization, Equation (7.94) shows that

$$\frac{g_L^2}{2} \sum_{\text{spins}} A^{\dagger} A = 32 g_L^2 s^2;$$

so the factor of two in the m^2 version of Equation (7.96), Equation (7.151) above, is correct. For the variable y, retaining the electron mass:

$$s+t+u = 2m^2; \ (p_1 \cdot p_2) = (s-m^2)/2; \ (p_2 \cdot p_3) = (m^2-u)/2;$$

$$y = 1 - \frac{p_2 \cdot p_3}{p_1 \cdot p_2} = 1 - \frac{m^2-u}{s-m^2} = -\frac{t}{s-m^2}.$$

Problem 7.6. Combined CC and NC cross-section for $\nu_e + e^- \to \nu_e + e^-$. Both W exchange and Z exchange are t channel, and we have written down both amplitudes: Equation (7.2) for charged current

$$\langle \nu_e e^- | T | \nu_e e^- \rangle$$
$$= \frac{G_F}{\sqrt{2}} \{ \bar{u}(p'_\nu) \gamma^\lambda (1-\gamma^5) u(p_e) \bar{u}(p'_e) \gamma_\lambda (1-\gamma^5) u(p_\nu) \},$$

$$(7.152)$$

and Equation (7.92) for the neutral current, written in the same notation:

$$\langle \nu_e e^- | T | \nu_e e^- \rangle$$
$$= \frac{G_F}{\sqrt{2}} \{ \bar{u}(p'_\nu) \gamma^\mu (1-\gamma^5) u(p_\nu) \bar{u}(p'_e) \gamma_\mu (g_V - g_A \gamma^5) u(p_e) \}.$$

$$(7.153)$$

Comparing these two formulas, apart from the different constants in the electron term for the neutral current, it is apparent that they have the same form if two of the spinors are interchanged, thus swap $\bar{u}(p'_\nu)$ and $\bar{u}(p'_e)$ in Equation (7.152). It is a feature of the algebra of the Dirac matrices that this swap can be done, with the price of a minus sign. In terms of components of the matrices:

$$(\gamma^\mu (1-\gamma^5))_{\alpha\beta} (\gamma_\mu (1-\gamma^5))_{\rho\sigma} = -(\gamma^\mu (1-\gamma^5))_{\rho\beta} (\gamma_\mu (1-\gamma^5))_{\alpha\sigma}.$$

$$(7.154)$$

This is called a Fierz transformation, after the theorist who first derived these formulas, and it is by no means obvious that you can do this. The transformation for $(V-A)$ is particularly simple, because it maps into $(V-A)$. The general reordering theorem involves all five invariants (S,V,T,A,P) and is a 5×5 matrix given in Appendix A of Barger and Phillips. Application of the reordering theorem to problems in Dirac algebra can simplify some calculations. So, we may rewrite Equation (7.152) as

$$\langle \nu_e e^- | T | \nu_e e^- \rangle$$

$$= -\frac{G_F}{\sqrt{2}} \{ \bar{u}(p_e')\gamma^\lambda (1 - \gamma^5) u(p_e) \bar{u}(p_\nu')\gamma_\lambda (1 - \gamma^5) u(p_\nu) \},$$

$$(7.155)$$

There is a bit of subtlety regarding the relative algebraic sign of the two diagrams to begin with. Two fermion lines are interchanged between the W and Z exchange diagrams, and that gives a relative minus sign, that is canceled by the Fierz transformation. So, the final answer is

$$\langle \nu_e e^- | T | \nu_e e^- \rangle = \frac{G_F}{\sqrt{2}} \{ \bar{u}(p_\nu')\gamma^\mu (1 - \gamma^5) u(p_\nu) \bar{u}(p_e')\gamma_\mu ((1 + g_V)$$

$$- (1 + g_A)\gamma^5) u(p_e) \} .$$

$$(7.156)$$

Obtain the cross-section for $\nu_e e^-$ scattering for the sum of CC and NC exchange.

Solution 7.6. This problem introduces the Fierz transformation, an interesting property of Dirac algebra, that can lead to simplifications if you know how to use it. We are asked to solve the problem for $\nu_e \, e^- \to \nu_e \, e^-$ scattering using both t channel W exchange (charged current), and t channel Z exchange (neutral current). The charged current amplitude is given by Equation (7.152):

$$\langle \nu_e e^- | T | \nu_e e^- \rangle = \frac{G_F}{\sqrt{2}} \left\{ \bar{u}(p_\nu')\gamma^\lambda (1 - \gamma^5) u(p_e) \bar{u}(p_e')\gamma_\lambda (1 - \gamma^5) u(p_\nu) \right\};$$

and by Equation (7.153) for the neutral current:

$$\langle \nu_e e^- | T | \nu_e e^- \rangle = \frac{G_F}{\sqrt{2}} \{ \bar{u}(p_\nu')\gamma^\mu (1 - \gamma^5) u(p_\nu) \bar{u}(p_e')$$

$$\times \gamma_\mu (g_V - g_A \gamma^5) u(p_e) \} .$$

As mentioned in the statement of the problem, these two amplitudes differ by the interchange of two spinors: $\bar{u}(p_\nu')$ and $\bar{u}(p_e')$. This swap is a permutation of the indices on the γ matrices, and is independent of the spinors themselves, so it works for positrons as well as electrons. Derivation is an exercise in Dirac algebra. The interested reader is referred to an online source: C.C. Nishi, "Simple derivation of general Fierz type identities" arXiv:hep-ph0412245v4 17 Jan 2006. The $V - A$ interaction $\gamma^\mu(1 - \gamma^5)$ requires special treatment. Nishi calls them "chiral states". The result is:

$$(\gamma^\mu(1 - \gamma^5))_{\alpha\beta}(\gamma_\mu(1 - \gamma^5))_{\rho\sigma} = -(\gamma^\mu(1 - \gamma^5)_{\rho\beta}(\gamma_\mu(1 - \gamma^5))_{\alpha\sigma}.$$

The first index has been switched, and that exchanges the two $\bar{u}$'s, with a change in sign. Under a Fierz transformation $V - A$ maps into itself. The minus sign cancels the minus that comes from the exchange of the two fermions in the final state, giving a plus, and Equation (7.156):

$$\langle \nu_e e^- | T | \nu_e e^- \rangle = \frac{G_F}{\sqrt{2}} \{ \bar{u}(p'_\nu) \gamma^\mu (1 - \gamma^5) u(p_\nu) \bar{u}(p'_e) \gamma_\mu ((1 + g_V)$$
$$ - (1 + g_A) \gamma^5) u(p_e) \} .$$

Proceed with the trace evaluation:

$$\frac{1}{2} \sum_{\text{spins}} | \langle \nu_e e^- | T | \nu_e e^- \rangle |^2 = \frac{G_F^2}{4} N^{\mu\lambda} E_{\mu\lambda};$$

where the neutrino tensor is the same for charged and neutral currents:

$$N^{\mu\lambda} = Tr(\not{p}'_\nu \gamma^\mu (1 - \gamma^5) \not{p}_\nu \gamma^\lambda (1 - \gamma^5));$$

and the electron tensor has the neutral current couplings:

$$E_{\mu\lambda} = Tr(\not{p}'_e \gamma_\mu \not{p}_e \gamma_\lambda (1 + g_V) - \not{p}'_e \gamma_\mu \gamma^5 \not{p}_e \gamma_\lambda (1 + g_A)(1 + g_V - (1 + g_A)\gamma^5)).$$

The two tensors have the same form if $g_V = g_A = 0$ (no neutral current). The neutrino tensor has been evaluated in the text:

$$N^{\mu\lambda} = 8(p'^\mu_\nu p^\lambda_\nu + p'^\lambda_\nu p^\mu_\nu - g^{\mu\lambda}(p'_\nu \cdot p_\nu) - i\epsilon^{\alpha\mu\beta\lambda} p'_{\nu\alpha} p_{\nu\beta}).$$

The electron tensor is the same, except for the neutral current coupling constants:

$$E_{\mu\lambda} = 4(((1 + g_V)^2 + (1 + g_A)^2)(p'_{e\mu} p_{e\lambda} + p'_{e\lambda} p_{e\mu} - g_{\mu\lambda}(p'_e \cdot p_e))$$
$$ - 2i(1 + g_V)(1 + g_A)\epsilon_{\sigma\mu\tau\lambda} p'^\sigma_e p^\tau_e));$$

observe that these two tensors are indeed the same if $g_V = g_A = 0$. The contraction has the usual cancellations:

$$E_{\mu\lambda} N^{\mu\lambda} = 64(((1 + g_V)^2 + (1 + g_A)^2)((p'_e \cdot p'_\nu)(p_e \cdot p_\nu)$$
$$+ (p_e \cdot p'_\nu)(p'_e \cdot p_\nu))$$
$$+ 2(1 + g_V)(1 + g_A)((p'_e \cdot p'_\nu)(p_e \cdot p_\nu) - (p_e \cdot p'_\nu)(p'_e \cdot p_\nu))).$$

In terms of Mandelstam variables:

$$\frac{1}{2} \sum_{\text{spins}} |\langle \nu_e e^- | T | \nu_e e^- \rangle|^2 = 4G_F^2((2 + g_V + g_A)^2 s^2 + (g_V - g_A)^2 u^2)$$

$$= 4G_F^2 s^2 (2 + g_V + g_A)^2 + (g_V - g_A)^2 (1 - y)^2);$$

where $y = (u + s)/s$, or $-u/s = 1 - y = (1 + \cos\theta)/2$ in the center of mass. Changing the coupling to left- and right-handed helicity states- $g_L = (g_V + g_A)/2$, $g_R = (g_V - g_A)/2$ - gives a nice form that matches other neutral current neutrino cross-sections:

$$\frac{1}{2} \sum_{\text{spins}} |\langle \nu_e e^- | T | \nu_e e^- \rangle = 16G_F^2 s^2 ((1 + g_L)^2 + g_R^2 (1 - y)^2).$$

Integration over the solid angle in the center of mass gives a total cross-section proportional to:

$$\sigma(\nu_e e^- \to \nu_e e^-) \sim G_F^2 s^2 \left((1 + g_L)^2 + \frac{g_R^2}{3} \right).$$

The numbers from Table 7.2 are $g_L = -0.28$; $g_R = 0.22$, giving a cross-section that is $\sim 1/2$ the size of the charged current alone. The angular dependence comes from right-handed electrons that couple through the neutral current, but the multiplier $g_R^2 = 0.05$ is small.

Problem 7.7. Take Equation (7.65) for the decay amplitude $W^+ \to \mu^+ \nu_\mu$. Suppose the production mechanism was $u\bar{d} \to W^+$, so that the W^+ spin is along the $\bar{d}$ direction, opposite to the proton beam, with $m_s = -1$, or the polarization vector for the W is $\epsilon = (0, 1, -i, 0)/\sqrt{2}$. Evaluate the trace for this polarization vector, and show that the μ^+ angular distribution in the W rest frame has the form $(1 - \cos\theta)^2$, where θ is the polar angle between the μ^+ and the proton beam. The μ^+'s, being right handed, are emitted backwards. Equation (7.36) is the solution to both production and decay. This exercise involves manipulating a spin one polarization vector. For the neutrino, which is the lepton in the final state, the sign of the $\cos\theta$ term flips, confirming the rule that the lepton follows the quark (u in the proton).

Solution 7.7. W production and decay in $p\bar{p}$ collisions. Begin with Equation (7.65):

$$\langle\mu^+\nu_\mu|\,T\,|W^+\rangle = \frac{g}{2\sqrt{2}}\epsilon_\nu^\lambda\bar{u}(p_3)\gamma^\nu(1-\gamma^5)v(p_4).$$

Now square it and sum over lepton spins, but not the W spin vectors:

$$\sum_{\text{spins}}|\langle\mu\nu_\mu|T\,|W^+\rangle|^2 = \frac{G_F M_W^2}{\sqrt{2}}\epsilon_\nu^\lambda\epsilon_\mu^{*\lambda}Tr(\not p_3\gamma^\nu(1-\gamma^5)\not p_4\gamma^\mu(1-\gamma^5))$$

$$= \sqrt{2}G_F M_W^2\epsilon_\nu^\lambda\epsilon_\mu^{*\lambda}(Tr(\not p_3\gamma^\nu\not p_4\gamma^\mu)$$
$$-Tr(\not p_3\gamma^\nu\gamma^5\not p_4\gamma^\mu))$$
$$= 4\sqrt{2}G_F M_W^2\epsilon_\nu^\lambda\epsilon_\mu^{*\lambda}(g^{\alpha\nu}g^{\beta\mu}+g^{\alpha\mu}g^{\beta\nu}$$
$$-g^{\alpha\beta}g^{\mu\nu}-i\epsilon^{\alpha\nu\beta\mu})p_{3\alpha}p_{4\beta}.$$

We have used $g^2/(8M_W^2) = G_F/\sqrt{2}$. Note that the square of the matrix element is proportional to G_F, not G_F^2. This is characteristic of production or decay of W bosons. Choose $+z$ to be the proton (up quark) direction. Then as argued in the statement of this problem, and shown in the derivation of Equation (7.56), $\lambda = -1$ for the W. W production by $u\bar{d}$ annihilation and W decay can be calculated with the same transposed matrix element. So the W polarization vector is:

$$\epsilon = (0,1,-i,0)/\sqrt{2};\ \epsilon^* = (0,1,i,0)/\sqrt{2};\ \mu=\nu=1;\ \mu=\nu=2;$$
$$\nu=1,\mu=2;\ \nu=2,\mu=1.$$

There are four terms in the sum over μ and ν:

$$\sum_{\text{spins}}|\langle\mu\nu_\mu|T\,|W^+\rangle|^2 = \frac{G_F M_W^2}{\sqrt{2}}(2g^{\alpha1}g^{\beta1}+2g^{\alpha2}g^{\beta2}+2g^{\alpha\beta}$$
$$+\epsilon^{\alpha1\beta2}-\epsilon^{\alpha2\beta1})p_{3\alpha}p_{4\beta}$$

$$= \sqrt{2} G_F M_W^2 (p_3^1 p_4^1 + p_3^2 p_4^2 + (p_3 \cdot p_4)$$
$$+ \epsilon^{\alpha 1 \beta 2} p_{3\alpha} p_{4\beta})$$
$$= \sqrt{2} G_F M_W^2 (p_3^0 p_4^0 - p_3^3 p_4^3 + p_{30} p_{43} - p_{40} p_{33}).$$

p_4 is the μ^+, and $\vec{p}_4 \cdot \hat{z} = M_Z \cos\theta/2$. The lower index four vectors are the neutrino $p_{3\alpha} = M_W/2(1, \sin\theta, 0, \cos\theta)$; and the muon $p_{4\beta} = M_W/2(1, -\sin\theta, 0, -\cos\theta)$. Putting all this together gives:

$$\sum_{\text{spins}} |\langle \mu\nu_\mu | T | W^+ \rangle|^2 = \frac{\sqrt{2} G_F M_W^4}{4} (1 + \cos^2\theta - 2\cos\theta)$$

$$= \frac{\sqrt{2} G_F M_W^4}{4} (1 - \cos\theta)^2.$$

The lepton (the ν_μ) follows the quark (the u); the μ^+ likes to go backwards.

Problem 7.8. This problem requires access to a computer. Generate the angular distributions for μ^+ and μ^- relative to the proton beam direction for $p + \bar{p} \to W^\pm \to \mu^\pm + \nu_\mu$, for $\sqrt{s} = 2\,\text{TeV}$. Assume the parton model with no transverse momentum, so the $W^\pm$ are collinear with the beams. Calculate $y_{\max} = \ln((E_{\max} + p_{\max})/(E_{\max} - p_{\max}))/2$ for the W's. Now generate y_W flat between -2 and $+2$, and shift the distribution by one unit towards the protons for W^+ and the antiprotons for W^-. This is an approximate model of the statement that W^+'s tend to follow protons, and W^-'s antiprotons. Hop on the W rest frame, and pick the muon decay angle according to the distribution $(1 \pm \cos\theta)^2$. Calculate the muon rapidity in the W rest frame, and use the fact that rapidities add to obtain separately the $\mu^\pm$ rapidities in the lab. The polar angle distributions then follow from the fact that the muon transverse momentum is Lorentz invariant. There are many ways to generate distributions with random numbers. Several dimensions require sophisticated techniques, but one dimension is easy, and there are at least two ways to do it. If $f(x)$ is normalized:

$$\int_a^b dx\, f(x) = 1,$$

and R is a random number $0 \le R \le 1$. Then let

$$R = \int_a^x dy f(y),$$

and solve for x. x will be distributed according to $f(x)$. Since $f(x)$ is frequently a quadratic equation, as is the case for these angular distributions, this leads to solutions to cubic equations. For $f(x) = 3(1+x)^2/8$ the solution is particularly simple: $x = 2R^{1/3} - 1$. If $f(x)$ is not convenient to integrate, there is another way. Plot $f(x)$ vs. x in the range of interest, and normalize it so that for $a \le x \le b$ $f(x)_{\max} = 1$. Then pick $x_0 = (a - b)R_1 + b$, and compare $f(x_0)$ with a second random number R_2. If $f(x_0) > R_2$, the random number is inside the curve, keep the event, otherwise discard it and start over with a new value for R_1. This will also generate the curve, and is free of hard labor.

If you have access to PDF's, you can generate 'weighted events' by multiplying the flat W rapidity distribution with $-y_{\max} \le y_W \le +y_{\max}$ by $x_1 x_2 u(x_1) d(x_2)$ for W^+ and $x_1 x_2 d(x_1) u(x_2)$ for W^-, where $x_{1,2} = M_W e^{\pm y_W}/\sqrt{s}$. If you do this the W rapidity distribution will have the correct shape, and will be shifted towards the protons for W^+ and towards the antiprotons for W^-. You will probably need 100 times as many events to get comparable numbers. This has no effect on the decay in the W center of mass.

Solution 7.8. Model the parton distribution shift towards protons for W^+ and $\bar{p}$'s for W^- by shifting the rapidity distributions one unit plus and minus respectively. This may overestimate the effect. The muon angular distributions in the W rest frame are shown in Figure 2. The W rapidity cannot be measured. The only measured quantity is the $\mu^{\pm}$ rapidity in the laboratory, shown in Figure 3. The maximum W rapidity is $y_{\max} = \ln(\sqrt{s}/M_W) = 3.2$. In this problem we leave off the end points, and assume a flat distribution from $-2 \le y \le +2$. The curves were obtained using ROOT and its random number generator. Figure 4 shows the same thing, except that the EHLQ pdf's (given in Problem 5.12) were used to generate the W rapidity in the laboratory.

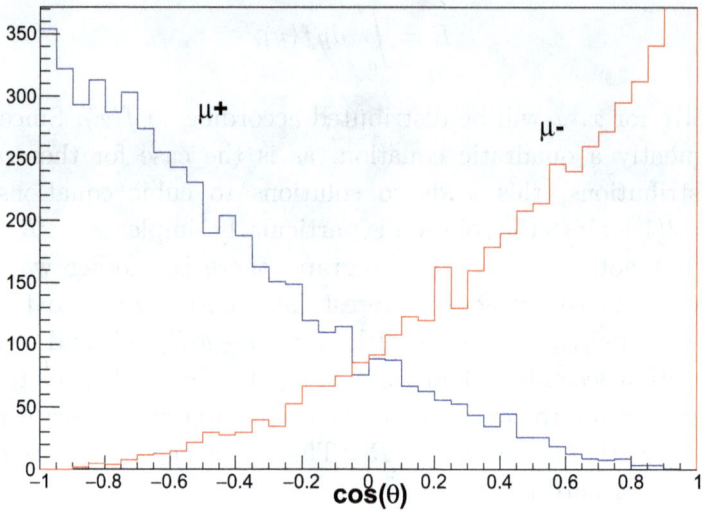

Figure 2: $\cos(\theta)$ distributions for μ^+ (blue, $(1 - \cos \theta)^2$), and μ^- (red $(1 + \cos \theta)^2$).$\theta = 0$ is in the proton direction.

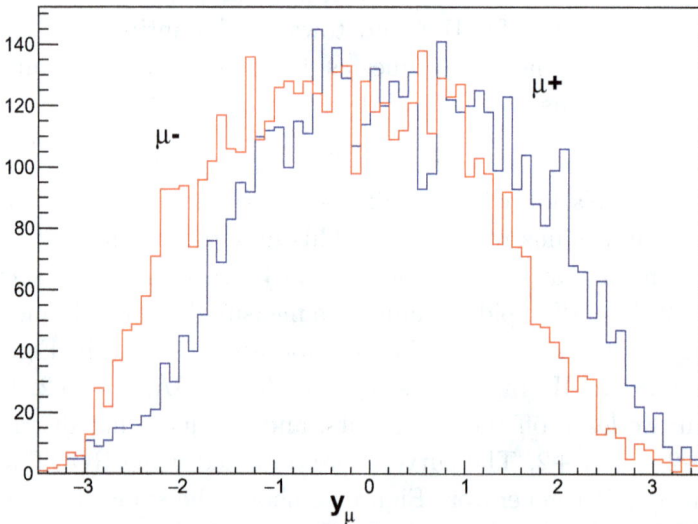

Figure 3: $\mu^{\pm}$ rapidity distributions in the laboratory. Note that for this choice of model of the W rapidity at production, the weak spin correlation is overwhelmed, and μ^+ is still shifted forward.

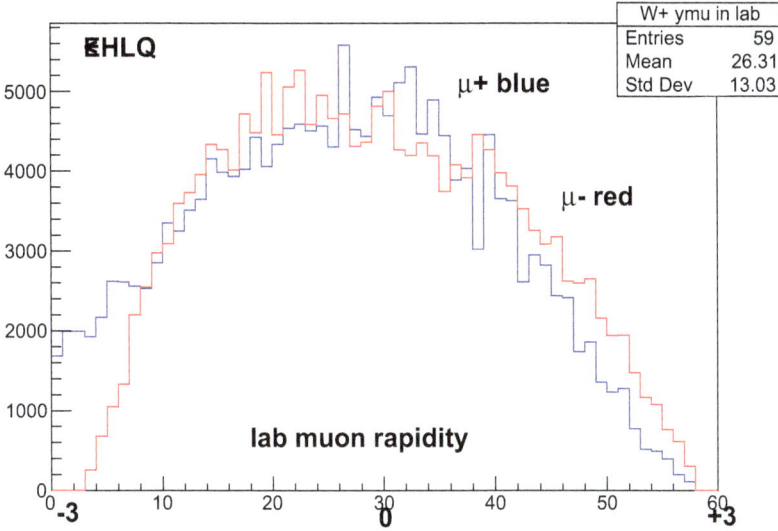

Figure 4: $\mu^{\pm}$ lab rapidity distributions using EHLQ pdf's and the parton model. In the central region near $y = 0$ the flow of charge, or the u and d quark x distributions, dominate, while in the forward and backward rapidities the weak force takes over, and the charge asymmetry changes sign. For $W \to \mu\nu$ only the muon rapidity can be measured.

Problem 7.9. Calculate the decay rate for $t \to W^+ + s$, starting from first principles, Equation (7.73). Use the appropriate CKM matrix element from Equation (7.70). What is the branching ratio for $\Gamma(t \to Ws)/\Gamma(t \to all)$? How would you search for this decay in a very large $t\bar{t}$ event sample?

Solution 7.9. Top quark decay $t \to W^+ + s$. Change the CKM matrix element in Equation (7.73):

$$-i\langle Ws|\,T\,|t\rangle = \frac{-igV_{ts}}{2\sqrt{2}}\epsilon_\mu^{*\lambda}\bar{u}(p_2)\gamma^\mu(1-\gamma^5)u(p_1);$$

with p_1 the top quark and p_2 the strange quark; $p_3 = p_1 - p_2$ is the W in the final state. Proceeding with the spin sum:

$$\frac{1}{2}\sum_{\text{spins}}|\langle Ws|\,T\,|t\rangle|^2 = \frac{G_F M_W^2 |V_{ts}|^2}{\sqrt{2}}\sum_\lambda \epsilon_\mu^{*\lambda}\epsilon_\nu^\lambda \text{Tr}(\not{p}_2\gamma^\mu(1-\gamma^5)\not{p}_1\gamma^\nu).$$

The W spin sum is:

$$\sum_\lambda \epsilon_\mu^{*\lambda}\epsilon_\nu^\lambda = -g_{\mu\nu} + \frac{(p_1 - p_2)_\mu(p_1 - p_2)_\nu}{M_W^2}.$$

In this case there is a contribution to the trace from the four momentum term in the W spin sum.

$$\frac{1}{2}\sum_{\text{spins}} |\langle Ws|\, T\, |t\rangle|^2 = \frac{G_F M_W^2 |V_{ts}|^2}{\sqrt{2}}(-\text{Tr}(\not{p}_2\gamma^\mu\not{p}_1\gamma_\mu)$$

$$+\frac{1}{M_W^2}(\text{Tr}(\not{p}_2\not{p}_1(1-\gamma^5)\not{p}_1\not{p}_1)$$

$$-\text{Tr}(\not{p}_2\not{p}_1(1-\gamma^5)\not{p}_1\not{p}_2).$$

Here $p_1 \cdot p_1 = m_t^2$, $p_2 \cdot p_2 = 0$, and $p_1 \cdot p_2 = m_t E_2 = (m_t^2 - M_W^2)/2$.

$$\frac{1}{2}\sum_{\text{spins}} |\langle Ws|\, T\, |t\rangle|^2 = \frac{G_F M_W^2 |V_{ts}|^2}{\sqrt{2}}(8p_1 \cdot p_2)\left(1 + \frac{m_t^2}{2M_W^2}\right)$$

$$= \frac{2G_F |V_{ts}|^2}{\sqrt{2}}(m_t^2 - M_W^2)(2M_W^2 + m_t^2);$$

in agreement with Equation (7.77). According to Equation (7.70) (the CKM matrix elements), $|V_{ts}| = 0.041$, so $|V_{ts}|^2 = 0.0016$, making the branching fraction to strange quarks about $1/1000$. The world sample of $t\bar{t}$ pairs is probably large enough to justify a search for direct K mesons. The K would have to come directly from the W decay vertex, and not as a daughter from the B meson decay secondary vertex.

Problem 7.10. The helicity rules do not hold for $p\bar{p} \to Z$ production, because of the right-handed quarks, but left-handed still prevails, and although for $Z \to \mu^+\mu^-$ the asymmetry is small, it still favors left-handed fermions, so the rule that the lepton follows the quark holds for both W's and Z's — just not as convincing for Z's. Take Equation (7.130) and leave the polarization vector $\epsilon_\mu^{*\lambda}$ in the calculation to repeat the analysis done for W's beginning with Equation (7.49). Remember to lower the indices for the quark four momenta! You should obtain an expression like:

$$\sum_{\text{quark spins}} |\langle Z|T|u\bar{u}\rangle|^2 = \frac{2g^2}{\cos^2\theta_W}\{(g_L^2 + g_R^2)(p_1 \cdot \epsilon^\lambda p_2 \cdot \epsilon^{\lambda*}$$

$$+ p_1 \cdot \epsilon^{\lambda*} p_2 \cdot \epsilon^\lambda + p_1 \cdot p_2)$$

$$+ (g_R^2 - g_L^2)i\epsilon^{\alpha\mu\beta\nu}\epsilon_\mu^{\lambda*}\epsilon_\nu^\lambda p_{2\alpha} p_{1\beta}\}.$$

$$(7.157)$$

Now use the polarization vector components defined by Equation (7.25) to obtain:

$$\sum_{\text{quark spins}} |\langle Z|T|u\bar{u}\rangle|^2 = \begin{cases} \dfrac{16G_F}{\sqrt{2}}g_R^2 M_Z^4, \ \lambda = +1; \\[2ex] \dfrac{16G_F}{\sqrt{2}}g_L^2 M_Z^4, \ \lambda = -1; \\[2ex] 0, \ \lambda = 0. \end{cases} \qquad (7.158)$$

The sum of these two terms divided by four for initial spins agrees with Equation (7.133). Calculate the Z polarization for $u\bar{u}$ and $d\bar{d}$ production in $p\bar{p}$ collisions, using the numbers in Table 7.2.

Solution 7.10. Z polarization in $p\bar{p} \to Z$ in the parton model. Begin with Equation (7.130):

$$-i\langle Z|T|u\bar{u}\rangle = \frac{-ig}{2\cos\theta_w}\epsilon_\mu^{*\lambda}\bar{v}(p_2)\gamma^\mu(g_L^u(1-\gamma^5) + g_R^u(1+\gamma^5))u(p_1);$$

where p_1 is the u quark and p_2 is the $\bar{u}$ quark; $p_1 + p_2 = p_3$ is the Z. From Table 7.2 $g_L^u = 0.35$; $g_R^u = 0.15$; $g_L^d = -0.425$; $g_R^d = 0.074$. Here the cross terms vanish because $(1-\gamma^5)(1+\gamma^5) = 0$, giving:

$$\sum_{\text{spins}} |\langle Z|T|u\bar{u}\rangle|^2 = \frac{g^2}{2\cos^2\theta_W}\sum_\lambda \epsilon_\mu^{*\lambda}\epsilon_\nu^\lambda(((g_L^u)^2 + (g_R^u)^2)Tr(\not{p}_2\gamma^\mu\not{p}_1\gamma^\nu)$$

$$+ ((g_R^u)^2 - (g_L^u)^2)Tr(\not{p}_2\gamma^\mu\gamma^5\not{p}_1\gamma^\nu)).$$

If we do the spin sum, we use the completeness relation:

$$\sum_\lambda \epsilon_\mu^{*\lambda}\epsilon_\nu^\lambda = -g_{\mu\nu} + \frac{1}{M_Z^2}(p_{1\mu}p_{1\nu} + p_{2\mu}p_{2\nu} + p_{1\mu}p_{2\nu} + p_{1\nu}p_{2\mu}).$$

All of the terms multiplied by $1/M_Z^2$ vanish because $p_1 \cdot p_1 = p_2 \cdot p_2 = 0$, so we are left with the metric tensor:

$$= \frac{g^2}{2\cos^2\theta_W}(((g_L^u)^2 + (g_R^u)^2)(-Tr(\not{p}_2\gamma^\mu\not{p}_1\gamma_\mu))$$
$$+ ((g_R^u)^2 - (g_L^u)^2)(-Tr(\not{p}_2\gamma^\mu\not{p}_1\gamma_\mu\gamma^5))).$$

The γ^5 trace vanishes, leaving:

$$\sum_{\text{spins}} |\langle Z|T|u\bar{u}\rangle|^2 = \frac{2g^2}{\cos^2\theta_W}((g_L^u)^2 + (g_R^u)^2)M_Z^2;$$

where we have used $p_1 \cdot p_2 = s/2 = M_Z^2/2$.

The next exercise is to back up, and not sum over λ. The lower index vectors in the Z rest frame are:

$$p_1 = (E, 0, 0, -E); \ p_2 = (E, 0, 0, E); \text{ where}$$
$$E = M_Z/2; \ \epsilon^0 = (0, 0, 0, -1);$$
$$\epsilon^{+1} = (0, 1, i, 0)/\sqrt{2}; \text{ and } \epsilon^{-1} = (0, -1, i, 0)/\sqrt{2}.$$

We are working with the formula:

$$\frac{g^2}{2\cos^2\theta_W}\epsilon_\mu^{*\lambda}\epsilon_\nu^\lambda(((g_L^u)^2 + (g_R^u)^2)Tr(\not{p}_2\gamma^\mu\not{p}_1\gamma^\nu)$$
$$+ ((g_R^u)^2 - (g_L^u)^2)Tr(\not{p}_2\gamma^\mu\not{p}_1\gamma^\nu\gamma^5)).$$

For $\lambda = 0, \mu = \nu = 3$, the traces vanish. For $\lambda = \pm 1$ we have

	$\mu = 1; \nu = 1$	$\mu = 1; \nu = 2$	$\mu = 2; \nu = 1$	$\mu = 2; \nu = 2$
$\lambda = +1; \ \epsilon\epsilon* =$	$1/2$	$i/2$	$-i/2$	$1/2$
$\lambda = -1; \ \epsilon\epsilon* =$	$1/2$	$-i/2$	$i/2$	$1/2$

So for $\lambda = +1$:

$$\frac{g^2}{2\cos^2\theta_W}\frac{1}{2}(((g_L^u)^2 + (g_R^u)^2)(Tr(\not{p}_2\gamma^1\not{p}_1\gamma^1) + Tr(\not{p}_2\gamma^2\not{p}_1\gamma^2)$$
$$- iTr(\not{p}_2\gamma^2\not{p}_1\gamma^1) + iTr(\not{p}_2\gamma^1\not{p}_1\gamma^2))$$
$$+ ((g_R^u)^2 - (g_L^u)^2)(-iTr(\not{p}_2\gamma^2\not{p}_1\gamma^1\gamma^5) + iTr(\not{p}_2\gamma^1\not{p}_1\gamma^2\gamma^5))).$$

The imaginary terms vanish because p_1 and p_2 have no x, y components.

$$= \frac{2g^2}{\cos^2 \theta_W}(((g_L^u)^2 + (g_R^u)^2)M_Z^2/2 - ((g_R^u)^2 - (g_L^u)^2)$$

$$\times (\epsilon^{0231} p_{20}p_{13} + \epsilon^{3201} p_{23}p_{10}))$$

$$= \frac{g^2 M_Z^2}{\cos^2 \theta_W}(((g_L^u)^2 + (g_R^u)^2) - ((g_R^u)^2 - (g_L^u)^2)\epsilon^{0231})$$

$$= 2\frac{g^2 M_Z^2}{\cos^2 \theta_W}(g_R^u)^2.$$

This is the same as Equation (7.158), using $\cos^2 \theta_W = M_W^2/M_Z^2$ and $G_F/\sqrt{2} = g^2/(8M_W^2)$. For $\lambda = -1$, μ and ν are interchanged, which flips the sign of $\epsilon^{\alpha\mu\beta\nu}$, and replaces $(g_R^u)^2$ with $(g_L^u)^2$. The polarization is $P_Z = (g_R^2 - g_L^2)/(g_R^2 + g_L^2) = -0.7$ for u quarks and $P_Z = -0.94$ for d quarks. In each case the Z spin is pointing in the antiproton direction.

Problem 7.11. Show that the angular distribution of the final state fermion f for $Z \to f\bar{f}$ with the Z polarization state $\lambda = -1$ in the Z rest frame is

$$\sum_{f\bar{f} \text{ spins}} |\langle f\bar{f}|T|Z\rangle|^2 = \frac{g^2 M_Z^2}{2\cos^2 \theta_W}\left\{ (g_L^f)^2(1 + \cos\theta)^2 \right.$$

$$\left. + (g_R^f)^2(1 - \cos\theta)^2 \right\}. \tag{7.159}$$

θ is the polar angle of the outgoing fermion, and since the initial spin $m_s = -1$, the left-handed fermion likes to go forwards.

Solution 7.11. This is the companion to the previous problem, looking at the decay of a polarized Z. The initial and final states are interchanged. This means that the polarization vector is ϵ instead of ϵ^*, and the fermion four momenta are now functions of the angle, and that gives the angle dependence to the Z decay. The square of the transition matrix element summed over fermion spins is:

$$\sum_{f\bar{f}\text{spins}} |\langle f\bar{f}|T|Z\rangle|^2 = \frac{g^2}{2\cos^2\theta_W}\epsilon_\mu^\lambda\epsilon_\nu^{*\lambda}Tr(\not p_3\gamma^\mu((g_L^f)^2(1-\gamma^5)$$

$$+(g_R^f)^2(1+\gamma^5))\not p_4\gamma^\nu).$$

Here $p_3 = M_Z/2(1,-\sin\theta,0,-\cos\theta)$ is the f lower index four momentum, and $p_4 = M_Z/2(1,\sin\theta,0,\cos\theta)$ is the same for the $\bar{f}$. We are asked to evaluate this expression for Z spin along $-\hat{z}$, or $\lambda = -1$, for which the four vector is $\epsilon^{-1} = (0,-1,i,0)/\sqrt{2}$. The left- and right-handed terms differ only by the sign of γ^5, so we will outline the left part in detail. Summing over μ and ν gives:

$$\frac{g^2}{4\cos^2\theta_W}(g_L^f)^2(Tr(\not p_3\gamma^1(1-\gamma^5)\not p_4\gamma^1) + Tr(\not p_3\gamma^2(1-\gamma^5)\not p_4\gamma^2)$$

$$-iTr(\not p_3\gamma^1(1-\gamma^5)\not p_4\gamma^2) + iTr(\not p_3\gamma^2(1-\gamma^5)\not p_4\gamma^1))$$

$$= \frac{g^2}{4\cos^2\theta_W}(g_L^f)^2(Tr(\not p_3\gamma^1\not p_4\gamma^1) + Tr(\not p_3\gamma^2\not p_4\gamma^2)$$

$$+iTr(\not p_3\gamma^1\gamma^5\not p_4\gamma^2) - iTr(\not p_3\gamma^2\gamma^5\not p_4\gamma^1)).$$

The γ^5 terms with $\mu = \nu$ vanish. The traces may be evaluated by the usual procedure. The two γ^5 parts with $\mu \neq \nu$ combine, and of course $g^{11} = g^{22} = -1$:

$$= \frac{g^2}{\cos^2\theta_W}(g_L^f)^2(2g^{\alpha 1}g^{\beta 1} + g^{\alpha\beta} + 2g^{\alpha 2}g^{\beta 2} + g^{\alpha\beta} - 2\epsilon^{\alpha 1\beta 2})p_{3\alpha}p_{4\beta}$$

$$= \frac{g^2}{\cos^2\theta_W}(g_L^f)^2(2p_3^1p_4^1 + 2(p_3\cdot p_4) - 2\epsilon^{0132}p_{30}p_{43} + 2\epsilon^{3102}p_{33}p_{40})$$

$$= \frac{2g^2}{\cos^2\theta_W}(g_L^f)^2(p_{30}p_4^0 + p_3^3p_{43} - \epsilon^{0132}p_{30}p_{43} + \epsilon^{3102}p_{33}p_{40})$$

$$= \frac{g^2 M_Z^2}{2\cos^2\theta_W}(g_L^f)^2(1-\cos\theta)^2.$$

The sign of $\cos\theta$ comes from γ^5, so the $(g_R^f)^2$ term angular dependence is $(1+\cos\theta)^2$, in agreement with Equation (7.159). From Table 7.2 for the charged leptons $(g_R)^2 \sim (g_L)^2$, so the charge asymmetry

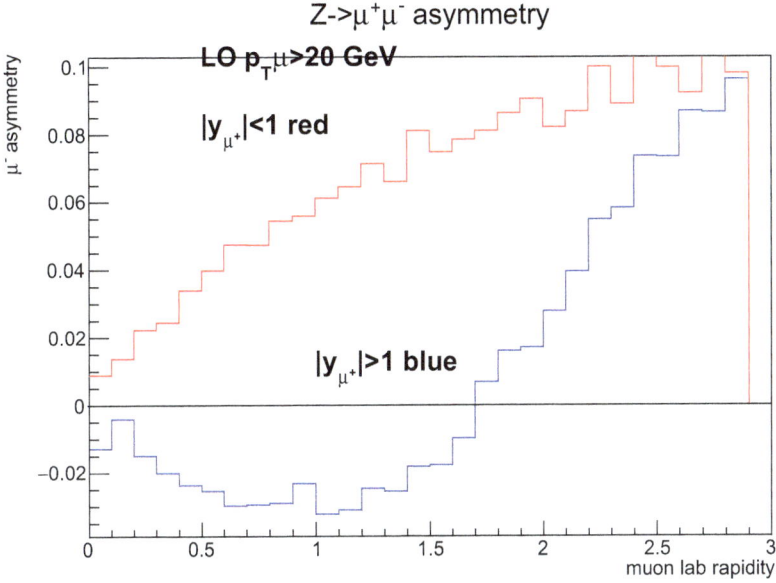

Figure 5: μ^- asymmetry as a function of rapidity, defined by $A = (N(y+) - N(y-))/(N(y+) + N(y-))$. The process is $p\bar{p} \to Z \to \mu^-\mu^+$ at 2 TeV. Each muon $p_T > 20$ GeV, a trigger requirement. For the red curve, $|y_{\mu^+}| < 1$ and for the blue curve $|y_{\mu^+}| > 1$. High rapidity μ^+ are predominantly backwards, and the motion of the Z pulls the μ^- backwards as well. A curious effect. These curves were produced with the parton model, using CTEQ5 pdf's.

is small. See Figure 5 for a plot of a parton model calculation of the μ^- asymmetry. For the quarks $(g_R)^2 << (g_L)^2$, and for the heavy quarks, where flavor and particle/antiparticle can be identified, this asymmetry has been measured. If $(g_R)^2 = (g_L)^2$, the angular distribution is $(1 + \cos^2 \theta)$, just like $e^- e^+ \to \mu^- \mu^+$ with no Z boson.

Chapter 8

Problem 8.1. Derive the commutation rule for two covariant derivatives in QED, Equation (3.207):

$$[D_\mu, D_\nu] = -ieF_{\mu\nu}. \tag{8.111}$$

Author's comments: This problem set is the last one to have typed up solutions. I did Chapter 9 earlier. I have to say that the subject matter of Chapter 8 is difficult. Non-Abelian gauge theory is not easy to handle. There is a sign error in Equation (8.25); the middle term should be a +. The Higgs mechanism, while algebraically straightforward, is conceptually mysterious. I have tried to limit the exposure in order to obtain important experimental predictions regarding the electroweak couplings and the Higgs boson itself. With that caveat we soldier on.

Solution 8.1. Commutator of covariant derivatives for QED:

$$D_\mu = \partial_\mu - ieA_\mu; \ [D_\mu, D_\nu]f(x) = ((\partial_\mu - ieA_\mu)(\partial_\nu - ieA_\nu)$$
$$-(\partial_\nu - ieA_\nu)(\partial_\mu - ieA_\mu))f(x)$$
$$= -ie(\partial_\mu A_\nu + A_\mu \partial_\nu - \partial_\nu A_\mu - A_\nu \partial_\mu)f(x)$$
$$= -ie(\partial_\mu(A_\nu) - \partial_\nu(A_\mu))f(x) = -ieF_{\mu\nu}f(x)$$

Problem 8.2. Repeat the calculation for the Yang–Mills covariant derivative to obtain Equation (8.16).

Solution 8.2. Commutator of covariant derivatives for Yang–Mills:

$$D_\mu = \partial_\mu + igW_{i\mu}T_i; \ [D_\mu, D_\nu]\psi(x)$$
$$= (ig(W_{i\mu}T_i\partial_\nu + \partial_\mu W_{i\nu}T_i - W_{i\nu}T_i\partial_\mu - \partial_\nu W_{i\mu}T_i)$$
$$-g^2 W_{i\mu}W_{k\nu}T_iT_k + g^2 W_{k\nu}W_{i\mu}T_kT_i)\psi(x)$$
$$= (ig(\partial_\mu W_{i\nu} - \partial_\nu W_{i\mu})T_i + g^2(T_kT_i - T_iT_k)W_{i\mu}W_{k\nu}))\psi(x).$$

Change the dummy index and use the commutation rule $[T_k, T_i] = i\epsilon_{kij}T_j$ to obtain:

$$[D_\mu, D_\nu] = ig(\partial_\mu W_{j\nu} - \partial_\nu W_{j\mu} + g\epsilon_{kij}W_{i\mu}W_{k\nu})T_j.$$

This agrees with Equations (8.14) and (8.15) because $\epsilon_{kij} = -\epsilon_{jik}$.

Problem 8.3. Derive Equation (8.18), proving that the Dirac equation is gauge invariant in the Yang–Mills theory.

Solution 8.3. Gauge invariance of the Dirac equation in the Lagrangian. We have to show that:

$$D'_\mu(1 - ig\chi_iT_i) = (1 - ig\chi_iT_i)D_\mu; \text{ where } D_\mu = \partial_\mu + igW_{i\mu}T_i; \text{ and}$$
$$D'_\mu = D_\mu + ig((\partial_\mu\chi_i) + g\epsilon_{ijk}\chi_jW_{k\mu})T_i.$$

We have put parentheses around $(\partial_\mu\chi_i)$ in the gauge transformed covariant derivative to remind ourselves that this is a vector function, and not an operator. We need to show that:

$$ig((\partial_\mu\chi_i) + g\epsilon_{ijk}\chi_jW_{k\mu})T_i - ig(\partial_\mu + igW_{k\mu}T_k)\chi_iT_i$$
$$= -ig\chi_iT_i(\partial_\mu + igW_{k\mu}T_k);$$

the terms multiplied by ig cancel:

$$ig((\partial_\mu\chi_i) - \partial_\mu\chi_i + \chi_i\partial_\mu)T_i = 0;$$

because there are two parts to $\partial_\mu\chi_i$ in the middle: $\partial_\mu\chi_i = (\partial_\mu\chi_i) + \chi_i\partial_\mu$. That leaves us with the g^2 terms, that vanish because of the commutation rules:

$$g^2(i\epsilon_{lik}T_l + T_kT_i - T_iT_k)\chi_iW_{k\mu} = 0.$$

This is true, because $[T_k, T_i] = i\epsilon_{kil}T_l$, and $\epsilon_{lik} = -\epsilon_{kil}$.

Problem 8.4. Prove that the free field Lagrangian Equation (8.14) is gauge invariant by deriving Equation (8.21) and demonstrating that Equation (8.22) proves invariance.

Solution 8.4. Gauge invariance of the free field Lagrangian. The double index field tensor is defined by Equation (8.14):

$$W_{i\mu\nu} = \partial_\mu W_{i\nu} - \partial_\nu W_{i\mu} - g\epsilon_{ijk}W_{j\mu}W_{k\nu}.$$

Equation (8.21) claims that the gauge transformation of $W_{i\mu\nu}$ is:

$$W_{i\mu\nu} \rightarrow W_{i\mu\nu} + g\epsilon_{ijk}\chi_j W_{k\mu\nu}.$$

The free field Lagrangian is:

$$\mathcal{L} = -W_{i\mu\nu}W_i^{\mu\nu}/4.$$

Using Equation (8.21), the Lagrangian is invariant if:

$$\epsilon_{ijk}\chi_j W_k^{\mu\nu}W_{i\mu\nu} + \epsilon_{ijk}\chi_j W_{k\mu\nu}W_i^{\mu\nu} = 0.$$

This vanishes because $\epsilon_{ijk} = -\epsilon_{kji}$. The next task is to show that Equation (8.21) is true. Gauge transform $W_{i\mu}$ and substitute into the tensor equation:

$$W_{i\mu} \rightarrow W_{i\mu} + (\partial_\mu\chi_i) + g\epsilon_{ijk}\chi_j W_{k\mu};\ W_{i\mu\nu}$$
$$\rightarrow \partial_\mu W_{i\nu} + \partial_\mu(\partial_\nu\chi_i) + g\epsilon_{ijk}\partial_\mu(\chi_j W_{k\nu})$$
$$- \partial_\nu W_{i\mu} - \partial_\nu(\partial_\mu\chi_i) - g\epsilon_{ijk}\partial_\nu(\chi_j W_{k\mu})$$
$$- g(\epsilon_{ijk}W_{j\mu} + \epsilon_{ijk}(\partial_\mu\chi_j) + g\epsilon_{ijk}\epsilon_{jlm}\chi_l W_{m\mu})$$
$$\times (W_{k\nu} + (\partial_\nu\chi_k) + g\epsilon_{knp}\chi_n W_{p\nu})$$
$$= W_{i\mu\nu} + g\epsilon_{ijk}\chi_j(\partial_\mu W_{k\nu} - \partial_\nu W_{k\mu})$$
$$- g^2\epsilon_{ijk}(\epsilon_{knp}\chi_n W_{j\mu}W_{p\nu} + \epsilon_{jlm}\chi_l W_{m\mu}W_{k\nu}).$$

Using the sum rule $\epsilon_{ijk}\epsilon_{ilm} = \delta_{jl}\delta_{km} - \delta_{jm}\delta_{kl}$ this expression reduces to Equation (8.23):

$$W_{i\mu\nu} \rightarrow W_{i\mu\nu} + g\epsilon_{ijk}\chi_j(\partial_\mu W_{k\nu} - \partial_\nu W_{k\mu}) + g^2(\vec{W}_\mu \cdot \vec{\chi}W_{i\nu} - \vec{W}_\nu \cdot \vec{\chi}W_{i\mu}).$$

That this equation and Equation (8.21) are the same can be shown by substituting $W_{k\mu\nu} = \partial_\mu W_{k\nu} - \partial_\nu W_{k\mu} - g\epsilon_{klm}W_{l\mu}W_{m\nu}$ into Equation (8.21) and using the ϵ theorem.

Problem 8.5. Obtain the Yang–Mills wave Equation (8.25) by taking the necessary functional derivatives of the Lagrangian: Equation (8.16) in Equation (8.26).

Solution 8.5. Yang–Mills wave equation. Write the part of the Lagrangian that involves derivatives according to the prescription of Equation (8.6):

$$-\frac{1}{4}\Big\{g^{\alpha\mu}g^{\beta\nu}(\partial_\mu W_{i\nu}\partial_\alpha W_{i\beta} - \partial_\mu W_{i\nu}\partial_\beta W_{i\alpha}$$
$$-\partial_\nu W_{i\mu}\partial_\alpha W_{i\beta} + \partial_\nu W_{i\mu}\partial_\beta W_{i\alpha})$$
$$-g\epsilon_{ijk}W_{j\mu}W_{k\nu}g^{\mu\rho}g^{\nu\lambda}(\partial_\rho W_{i\lambda} - \partial_\lambda W_{i\rho})$$
$$-g\epsilon_{ilm}g^{\mu\tau}g^{\nu\delta}(\partial_\mu W_{i\nu} - \partial_\nu W_{i\mu})W_{l\tau}W_{m\delta}\Big\}.$$

Now take the derivative with respect to $\partial_\gamma W_{i\sigma}$. You get Kronecker δ's everywhere, and the terms combine to cancel the $1/4$, just like Equation (8.6). The result is:

$$\frac{\partial\mathcal{L}}{\partial(\partial_\gamma W_{i\sigma})} = -(\partial^\gamma W_i^\sigma - \partial^\sigma W_i^\gamma - g\epsilon_{ijk}W_j^\gamma W_k^\sigma) = -W_i^{\gamma\sigma}.$$

The derivative $\partial/\partial x^\gamma$ gives the first term in Equation (8.25). For the other two terms we need the parts of $\mathcal{L}$ that depend on $W_{i\mu}$. They are:

$$-\frac{1}{4}\Big\{-g\epsilon_{ijk}W_{j\mu}W_{k\nu}(\partial^\mu W_i^\nu - \partial^\nu W_i^\mu)$$
$$- g\epsilon_{ilm}(\partial_\mu W_{i\nu} - \partial_\nu W_{i\mu})W_l^\mu W_m^\nu + g^2\epsilon_{ijk}\epsilon_{ilm}W_{j\mu}W_{k\nu}W_l^\mu W_m^\nu\Big\}$$
$$+ \bar{\psi}(i\gamma^\mu(\partial_\mu + igW_{i\mu}T_i) - m)\psi.$$

The two g terms can be combined by relabeling the indices:

$$\frac{g}{2}\frac{\partial}{\partial W_{n\sigma}}(\epsilon_{ijk}W_{j\mu}W_{k\nu}(\partial^\mu W_i^\nu - \partial^\nu W_i^\mu))$$
$$= \frac{g}{2}\epsilon_{ijk}(\delta_{jn}\delta_{\mu\sigma}W_{k\nu} + \delta_{kn}\delta_{\nu\sigma}W_{j\mu})(\partial^\mu W_i^\nu - \partial^\nu W_i^\mu)$$
$$= \frac{g}{2}(\epsilon_{ink}W_{k\nu}(\partial^\sigma W_i^\nu - \partial^\nu W_i^\sigma) + \epsilon_{ijn}W_{j\mu}(\partial^\mu W_i^\sigma - \partial^\sigma W_i^\mu)).$$

Changing dummy indices and turning the antisymmetric derivatives around combines the two expressions into one:

$$g\epsilon_{ink}W_{k\nu}(\partial^\sigma W_i^\nu - \partial^\nu W_i^\sigma) = g\epsilon_{nki}W_{k\nu}(\partial^\sigma W_i^\nu - \partial^\nu W_i^\sigma).$$

This is the same as the first term in Equation (8.27). For the g^2 term we need two metric tensors:

$$-\frac{g^2}{4}\frac{\partial}{\partial W_{n\sigma}}\epsilon_{ijk}\epsilon_{ilm}W_{j\mu}W_{k\nu}g^{\alpha\mu}g^{\beta\nu}W_{l\alpha}W_{m\beta}$$

$$= -\frac{g^2}{4}\epsilon_{ijk}\epsilon_{ilm}(\delta_{jn}\delta_{\mu\sigma}W_{k\nu}W_l^\mu W_m^\nu + W_{j\mu}\delta_{kn}\delta_{\nu\sigma}W_l^\mu W_m^\nu$$

$$+ W_{j\mu}W_{k\nu}g^{\alpha\mu}\delta_{\alpha\sigma}\delta_{ln}W_m^\nu + W_{j\mu}W_{k\nu}g^{\beta\nu}W_l^\mu\delta_{nm}\delta_{\sigma\beta})$$

$$= -\frac{g^2}{4}(\epsilon_{ilm}(\epsilon_{ink}W_{k\nu}W_m^\nu W_l^\sigma + \epsilon_{ijn}W_{j\mu}W_l^\mu W_m^\sigma)$$

$$+ \epsilon_{ijk}(\epsilon_{inm}W_j^\sigma W_{k\nu}W_m^\nu + \epsilon_{iln}W_{j\mu}W_k^\sigma W_l^\mu))$$

$$= -g^2(\vec{W}_\nu\cdot\vec{W}^\nu W_n^\sigma - \vec{W}_\nu\cdot\vec{W}^\sigma W_n^\nu).$$

We have used $\epsilon_{ink}\epsilon_{ilm} = (\delta_{nl}\delta_{km} - \delta_{kl}\delta_{nm})$ etc. to give eight terms that combine to cancel the $1/4$ in front. This agrees with the g^2 term in Equation (8.27). The only problem is that the Euler–Lagrange equation, Equation (8.26), has a minus sign. There is a sign error in Equation (8.25); it should read:

$$-\partial_\nu W_i^{\nu\sigma} + g\epsilon_{ijk}W_{j\nu}W_k^{\nu\sigma} + g\bar\psi\gamma^\sigma T_i\psi = 0.$$

The signs on the first two terms in Equation (8.27) also have to be reversed. The author regrets the error.

Problem 8.6. Obtain Equation (8.44) from Equation (8.42), demonstrating that the Z couplings used in Chapter 7 are correct.

Solution 8.6. Obtain Equation (8.44) from Equation (8.42), which reads:

$$j_{\mu Z} = \frac{g}{\cos\theta_W}(\bar\psi_L\gamma_\mu(T_3 - \sin^2\theta_W Q)\psi_L - \bar\psi_R\gamma_\mu\sin^2\theta_W Q\psi_R).$$

The left- and right-handed spinors are defined by $\psi_L = (1 - \gamma^5)\psi/2$ and $\psi_R = (1 + \gamma^5)\psi/2$. γ^5 is hermitian: $\gamma^{5\dagger} = \gamma^5$. Therefore:

$$\psi_L = (1 - \gamma^5)\psi/2; \quad \bar{\psi}_L = \bar{\psi}(1 + \gamma^5)/2; \text{ and}$$
$$\psi_R = (1 + \gamma^5)\psi/2; \quad \bar{\psi}_R = \bar{\psi}(1 - \gamma^5)/2.$$

Hence $\bar{\psi}_L\psi_L = \bar{\psi}_R\psi_R = 0$, precluding a mass term in a left- (or right-) handed Lagrangian. The γ_μ turns γ^5 into $-\gamma^5$, which gives $(1 \pm \gamma^5)^2 = 2(1 \pm \gamma^5)$. Hence:

$$j_{\mu Z} = \frac{g}{4\cos\theta_W}(\bar{\psi}(1 + \gamma^5)\gamma_\mu(T_3 - \sin^2\theta_W Q)(1 - \gamma^5)\psi$$
$$-\bar{\psi}(1 - \gamma^5)\gamma_\mu \sin^2\theta_W Q(1 + \gamma^5)\psi$$
$$= \frac{g}{2\cos\theta_W}(\bar{\psi}\gamma_\mu(T_3 - \sin^2\theta_W Q)\psi - \bar{\psi}\gamma_\mu \sin^2\theta_W Q\psi$$
$$+\bar{\psi}\gamma^5\gamma_\mu(T_3 - \sin^2\theta_W Q)\psi + \bar{\psi}\gamma^5\gamma_\mu \sin^2\theta_W Q\psi)$$
$$= \frac{g}{2\cos\theta_W}(\bar{\psi}\gamma_\mu(T_3 - 2\sin^2\theta_W Q)\psi - \bar{\psi}\gamma_\mu\gamma^5 T_3\psi).$$

In agreement with Equation (8.44). To check the signs, for a neutrino $T_3 = +1/2$ and $Q = 0$ so the interaction is pure $\gamma_\mu(1 - \gamma^5)$, and that is correct. For an electron $T_3 = -1/2$ and $Q = -1$, so the vector coupling is $-1/2 + 2\sin^2\theta_W$ which vanishes for $\theta_W = 30^0$, and that is also correct.

Problem 8.7. Do the algebra to show that the Lagrangian Equation (8.53) for the σ, π model is correct.

Solution 8.7. Verify Equation (8.53):

$$\mathcal{L} = \frac{1}{2}(\partial_\mu\sigma\partial^\mu\sigma + \partial_\mu\pi\partial^\mu\pi - 2\mu^2\sigma^2 - \frac{\lambda}{2}(\sigma^4 + \pi^4)$$
$$-2\sqrt{\lambda}\mu(\sigma^3 + \pi^2\sigma) - \lambda\sigma^2\pi^2).$$

Start with Equation (8.46):

$$\mathcal{L} = \frac{1}{2}(\partial_\mu\phi_1\partial^\mu\phi_1 + \partial_\mu\phi_2\partial^\mu\phi_2 + \mu^2(\phi_1^2 + \phi_2^2) - \frac{\lambda}{2}(\phi_1^2 + \phi_2^2)^2).$$

The potential is a Mexican hat in (ϕ_1, ϕ_2) space, as shown in Figure 8.1 in the text:

$$V(\phi) = -\frac{\mu^2}{2}(\phi_1^2 + \phi_2^2) + \frac{\lambda}{4}(\phi_1^2 + \phi_2^2)^2; \text{ let } \phi_2 = 0;$$

$$V(\phi_1) = -\frac{\mu^2}{2}\phi_1^2 + \frac{\lambda}{4}\phi_1^4; \quad \frac{\partial V}{\partial \phi_1} = (-\mu^2 + \lambda\phi_1^2)\phi_1 = 0.$$

We have two solutions: $\phi_1 = 0$; or $\phi_1 = \mu/\sqrt{\lambda}$. The Lagrangian is invariant under rotations in (ϕ_1, ϕ_2) space, and choosing the solution $\phi_1 = \mu/\sqrt{\lambda} = v$ breaks the symmetry. v is called the vacuum expectation value. Expand the two fields about v : $\phi_1 = v + \sigma(x)$; $\phi_2 = \pi(x)$, and substitute these expansions into the potential. The algebra is simple. It is the interpretation of the result that is remarkable. After dropping the constant term, that has no effect on the Euler–Lagrange equations, we find that the σ field has acquired a mass $m_\sigma = \sqrt{2}\mu$, that the π field remains massless, and the σ and π fields are coupled together with a constant m_σ^2/v. The derivatives are not affected by the constant v. Equation (8.53) can be written as:

$$\mathcal{L} = \frac{1}{2}(\partial^\mu \sigma \partial_\mu \sigma + \partial^\mu \pi \partial_\mu \pi - m_\sigma^2 \sigma^2 - \frac{\lambda}{2}(\sigma^2 + \pi^2)^2 - \frac{m_\sigma^2}{v}(\sigma^2 + \pi^2)\sigma).$$

This form has many of the features of the more complicated expressions of electroweak theory. Breaking the symmetry has generated a mass for ϕ_1, but ϕ_2 remains massless — a Goldstone boson. Self-coupling for ϕ_1 and $\phi_1 - \phi_2$ couplings are proportional to the square of the boson mass, and inversely proportional to the vacuum expectation value.

Problem 8.8. Evaluate the expectation value of the square of the covariant derivative Equation (8.65) and obtain the vector boson masses Equation (8.69).

Solution 8.8. Equation (8.65) is the vacuum expectation value of the square of the covariant derivative, without the ∂^μ operators:

$$\frac{1}{2}\begin{pmatrix} 0 & v \end{pmatrix}\left\{ig(W_{+\mu}T_+ + W_{-\mu}T_-) + i\left(gW_{3\mu}T_3 + g'\frac{B_\mu}{2}\right)\right\}$$

$$\times \left\{ig(W_+^\mu T_+ + W_-^\mu T_-) + i\left(gW_3^\mu T_3 + g'\frac{B^\mu}{2}\right)\right\}\begin{pmatrix} 0 \\ v \end{pmatrix}.$$

The operators can change the state vector:

$$T_+ \begin{pmatrix} 0 \\ v \end{pmatrix} = \frac{1}{\sqrt{2}} \begin{pmatrix} v \\ 0 \end{pmatrix}; \; T_- \begin{pmatrix} 0 \\ v \end{pmatrix} = 0; \; T_3 \begin{pmatrix} 0 \\ v \end{pmatrix} = -\frac{1}{2} \begin{pmatrix} 0 \\ v \end{pmatrix}.$$

So Equation (8.65) becomes:

$$= \frac{1}{2} \begin{pmatrix} 0 & v \end{pmatrix} \left(ig(W_{+\mu}T_+ + W_{-\mu}T_-) + i\left(gW_{3\mu}T_3 + \frac{g'B_\mu}{2} \right) \right)$$

$$\times \left(\frac{igW_+^\mu}{\sqrt{2}} \begin{pmatrix} v \\ 0 \end{pmatrix} - \frac{igW_3^\mu}{2} \begin{pmatrix} 0 \\ v \end{pmatrix} + \frac{ig'B^\mu}{2} \begin{pmatrix} 0 \\ v \end{pmatrix} \right)$$

$$= \frac{1}{2} \begin{pmatrix} 0 & v \end{pmatrix} \left(+\frac{g^2 W_{+\mu}W_3^\mu}{2\sqrt{2}} \begin{pmatrix} v \\ 0 \end{pmatrix} - \frac{gg'}{2\sqrt{2}} W_{+\mu}B^\mu \begin{pmatrix} v \\ 0 \end{pmatrix} \right.$$

$$-\frac{g^2}{2} W_{-\mu}W_+^\mu \begin{pmatrix} 0 \\ v \end{pmatrix} - \frac{g^2}{2\sqrt{2}} W_+^\mu W_{3\mu} \begin{pmatrix} v \\ 0 \end{pmatrix}$$

$$-\frac{g^2}{4} W_3^\mu W_{3\mu} \begin{pmatrix} 0 \\ v \end{pmatrix} + \frac{gg'}{4} W_{3\mu}B^\mu \begin{pmatrix} 0 \\ v \end{pmatrix} - \frac{gg'}{2\sqrt{2}} B_\mu W_+^\mu \begin{pmatrix} v \\ 0 \end{pmatrix}$$

$$\left. +\frac{gg'}{4} W_3^\mu B_\mu \begin{pmatrix} 0 \\ v \end{pmatrix} - \frac{g'^2}{4} B_\mu B^\mu \begin{pmatrix} 0 \\ v \end{pmatrix} \right)$$

$$= \frac{v^2}{8} (-2g^2 W_{-\mu}W_+^\mu - (gW_3^\mu - g'B^\mu)(gW_{3\mu} - g'B_\mu));$$

in agreement with Equation (8.67). Equation (8.68) then gives the masses in Equation (8.69) directly.

Problem 8.9. Calculate $(D_\mu\phi)^\dagger D^\mu\phi$ and obtain the couplings of the $W^\pm$ and Z to the Higgs, Equation (8.77).

Solution 8.9. Start with Equation (8.75):

$$D_\mu\phi = (\partial_\mu + ig(W_{+\mu}T_+ + W_{-\mu}T_-)$$

$$+ igW_{3\mu}T_3 + ig'B_\mu/2)\frac{1}{\sqrt{2}} \begin{pmatrix} 0 \\ v + h(x) \end{pmatrix}.$$

$$= \frac{1}{\sqrt{2}} \begin{pmatrix} i\frac{g}{\sqrt{2}} W_{+\mu}(v + h(x)) \\ \partial_\mu h(x) - i\frac{gZ_\mu}{2\cos\theta_W}(v + h(x)) \end{pmatrix};$$

where v and $h(x)$ are both real, and $W_+^{\mu\dagger} = W_-^\mu$ (refer to Equation (8.30); W_1 and W_2 are real fields). So $(D_\mu\phi)^\dagger D^\mu\phi$ becomes:

$$(D_\mu\phi)^\dagger D^\mu\phi = \frac{1}{2}\partial_\mu h(x)\partial^\mu h(x) + \frac{g^2}{4}W_-^\mu W_{+\mu}(v+h(x))^2$$

$$+ \frac{g^2 Z_\mu Z^\mu}{8\cos^2\theta_W}(v+h(x))^2).$$

The mass terms are $M_W^2 = g^2 v^2/4$ and $M_Z^2 = M_W^2/(\cos^2\theta_W)$. The factor of two comes from two charged W's, and only one Z. The terms linear in the Higgs field $h(x)$ correspond to WWh and ZZh couplings:

$$\frac{g^2}{4}W_{-\mu}W_+^\mu 2vh + \frac{g^2}{8\cos^2\theta_W}Z_\mu Z^\mu 2vh = 2\frac{M_W^2}{v}W_+W_-h + \frac{M_Z^2}{v}Z^2h.$$

Problem 8.10. Begin with Equation (8.81) and calculate the decay rate in MeV for $h \to b\bar{b}$. Use the PDG value for $m_b = 4.2\,\text{GeV}$.

Solution 8.10. $h \to f\bar{f}$ Equation (8.81):

$$\langle f\bar{f}|\,T\,|h\rangle = (\sqrt{2}G_F)^{1/2}m_f\bar{u}(p_1)v(p_2);\ (p_1+p_2)^2 = s = m_h^2.$$

$$\sum_{\text{spins}}|\langle f\bar{f}|\,T\,|h\rangle|^2 = \sqrt{2}G_F m_f^2 Tr((\not{p}_1 + m_f)(\not{p}_2 - m_f))$$

$$= 4\sqrt{2}G_F m_f^2(p_1\cdot p_2 - m_f^2)$$

For a $b\bar{b}$ final state:

$$= 2\sqrt{2}G_F m_b^2(m_h^2 - 4m_b^2) = 2\sqrt{2}G_F m_b^2 m_h^2\left(1 - 4\frac{m_b^2}{m_h^2}\right).$$

In agreement with Equation (8.83). $m_b^2/m_h^2 \sim 10^{-3}$, but we will carry it along. Note that the decay rate is proportional to G_F, not G_F^2. The formula is:

$$d\Gamma = \frac{|\langle b\bar{b}|\,T\,|h\rangle|^2}{2m_h(2\pi)^2}\delta^4(p - p_1 - p_2)\frac{d^2 p_1}{2E_1}\frac{d^3 p_2}{2E_2}$$

$$= \frac{2\sqrt{2}G_F m_b^2 m_h^2}{8\pi^2 m_h} \left(1 - \frac{4m_b^2}{m_h^2}\right) \delta(m_h - E_1 - E_2) \frac{d^3 p_1}{4E_1 E_2}$$

$$= \frac{2\sqrt{2}G_F m_b^2 m_h^2}{16\pi^2 m_h^3} \left(1 - \frac{4m_b^2}{m_h^2}\right) \delta(E_1 - m_h/2) p_1 E_1 dE_1 d\Omega.$$

The product $p_1 E_1 = m_h^2/4\sqrt{1 - 4m_b^2/m_h^2}$; and $\int d\Omega = 4\pi$, so

$$\Gamma = \frac{G_F m_b^2 m_h}{4\pi\sqrt{2}} \left(1 - \frac{4m_b^2}{m_h^2}\right)^{3/2}.$$

Multiplying by three for color gives Equation (8.85). Substituting in $m_b = 4.2$ GeV, $m_h = 125$ GeV, and $G_F = 1.166 \times 10^{-5}$ GeV^{-2} gives $\Gamma = 4.4$ MeV.

Problem 8.11. Calculate the cross-section in the $e^- e^+$ center of mass for $e^- e^+ \to Z + h$ at $\sqrt{s} = 250$ GeV. Use the diagram of Figure 8.4 turned around, with $e^- e^+$ annihilation in the initial state to create a Z propagator that decays into $Z+h$. It is a two-body final state, which is simpler than $h \to WW^* \to W\mu\nu$. Use the appropriate coupling of the Z to electrons. Express your final answer in units of $R = (4\pi\alpha^2)/(3s)$. The Higgs cross-section is very small. At 250 GeV $R = 1.36$ pb, giving about 5 muon pairs per hour at 10^{33} luminosity. It is interesting to note that the same initial and final states could go via a Higgs propagator, which is $1/(p^2 - m_h^2 + i\epsilon)$: $e^- e^+ \to h^* \to h + Z$, except for the fact that the last step does not exist. The Z can radiate a Higgs, but the Higgs cannot radiate a Z and conserve angular momentum.

The amplitude is

$$\langle Zh| T |e^- e^+\rangle = \frac{g}{2\cos\theta_W} \frac{(\sqrt{2}G_F)^{1/2} M_Z^2 \epsilon_\nu^{*\lambda}}{s - M_Z^2}(-g^{\mu\nu})$$
$$\times \bar{v}(k')\gamma_\mu(g_L(1 - \gamma^5) + g_R(1 + \gamma^5))u(k).$$

$$(8.112)$$

A two-body final state in the center of mass has no free parameters other than the angles. The two momenta are equal and opposite, and the two energies are given by familiar formulas:

$$E_Z = \frac{s + M_Z^2 - m_h^2}{2\sqrt{s}}, \qquad E_h = \frac{s + m_h^2 - M_Z^2}{2\sqrt{s}}. \tag{8.113}$$

The spin averaged square of the amplitude is

$$\sum_{\text{spins}} |\langle Zh| T |e^- e^+\rangle|^2$$

$$= \frac{16 G_F^2 M_Z^6}{(s - M_Z^2)^2} (g_L^2 + g_R^2) \left(k \cdot k' + \frac{2}{M_Z^2} (k \cdot p')(k' \cdot p') \right), \tag{8.114}$$

where p' is the Z four momentum in the final state, and k and k' are the initial state four momenta for e^- and e^+. $s = (k + k')^2 = (p + p')^2$. Integrate the differential cross-section over the Z solid angle to obtain:

$$\sigma = \frac{2}{3\pi} \frac{G_F^2 M_Z^4 |\vec{p}|}{\sqrt{s}} (g_L^2 + g_R^2) \frac{E_Z^2 + 2M_Z^2}{(s - M_Z^2)^2}, \tag{8.115}$$

where $|\vec{p}|$ is the momentum of either particle in the final state. Obtain the coupling constants for eeZ from Table 7.2, and numerically evaluate the cross-section in units of R. I got a cross-section $\sigma \sim 0.25$ pb, less than one unit of R. This puts pressure on the storage ring or linear collider to achieve the highest possible luminosity.

Solution 8.11. Cross-section for $e^- e^+ \to Z + h$ at $\sqrt{s} = 250$ GeV. The solution to this problem is well outlined in the problem itself. Equation (8.112) comes from the diagram. $(\sqrt{2} G_F)^{1/2} M_Z^2 \epsilon_\nu^*$ comes from the Zh vertex, and $g(-g^{\mu\nu})/(2\cos\theta_W(s - M_Z^2))$ comes from the coupling of the Z propagator to the $e^- e^+$ pair.

$$\langle Zh| T |e^- e^+\rangle = \frac{g}{2\cos\theta_W} \frac{(\sqrt{2} G_F)^{1/2} M_Z^2 \epsilon_\nu^{*\lambda}}{s - M_Z^2} (-g^{\mu\nu}) \bar{v}(k') \gamma_\mu$$

$$\times (g_L(1 - \gamma^5) + g_R(1 + \gamma^5)) u(k).$$

The sum over polarization states for a massive spin one particle like the W or Z is given by Equation (7.27):

$$\sum_\lambda \epsilon_\mu^{\lambda*} \epsilon_\nu^\lambda = -g^{\mu\nu} + \frac{p'_\mu p'_\nu}{M_Z^2};$$

where p'_μ is the four momentum of the Z, and this extra term contributes to the trace, as shown by Equation (8.114):

$$\sum_{\text{spins}} |\langle Zh| T |e^- e^+\rangle|^2 = \frac{16 G_F^2 M_Z^6}{(s - M_Z^2)^2} (g_L^2 + g_R^2)(k\cdot k' + \frac{2}{M_Z^2}(k\cdot p')(k'\cdot p'));$$

where p' is the Z four momentum in the final state, and k and k' are the initial state four momenta for e^- and e^+. $s = (k + k')^2 = (p+p')^2$. Integrate the differential cross-section over the Z solid angle to obtain:

$$\sigma = \frac{2}{3\pi} \frac{G_F^2 M_Z^4 |\vec{p}\,'|}{\sqrt{s}} (g_L^2 + g_R^2) \frac{E_Z^2 + 2M_Z^2}{(s - M_Z^2)^2}.$$

For $\sqrt{s} = 250$ GeV; $E_Z = 110$ GeV; $|\vec{p}\,'| = 62$ GeV; and $(g_L^2 + g_R^2) = 0.125$ for coupling the Z to electrons.

$$\sigma = \frac{2 \times 1.34 \times 10^{-10} \times 6.8 \times 10^6 \times 62 \times 0.125 \times 2.9 \times 10^4}{3\pi \times 250 \times 2.9 \times 10^8}$$

$$= 6.1 \times 10^{-10} \text{GeV}^{-2}$$

$$= 2.4 \times 10^{-37} \text{cm}^2 = 0.24 \text{ pb}.$$

This is the same number as quoted in the problem. Thus $\sigma(e^- e^+ \to Z + h) = 0.2 \times R$ at $\sqrt{s} = 250$ GeV. This gives only about 15 events per day at 10^{33} cm^{-2} sec^{-1} luminosity for 100% detection efficiency, and places severe luminosity requirements on a practical $e^- e^+$ machine to study the Higgs.

This same diagram, Figure 8.4, can be used to calculate a Drell–Yan type process to produce $Z + h$ in $p - \bar{p}$ or $p - p$ collisions. Just replace the leptons with quarks. This channel contributes to Higgs production at the LHC. It has the advantage of a distinct $Z \to l^- l^+$ signature (with the cost of more than a factor of 10 reduction in effective cross-section).

Problem 8.12. W magnetic moment. Do the parts integration on Equation (8.106) to obtain the two identical contributions to Equation (8.109). Then check that $\vec{\epsilon}^{\,\lambda=1} * \times \vec{\epsilon}^{\,\lambda=1} = \hat{z}$.

Solution 8.12. W magnetic moment. Equation (8.106) reads:

$$\frac{1}{2} g \epsilon_{ijk} (\partial_\mu W_{i\nu} - \partial_\nu W_{i\mu}) W_j^\mu W_k^\nu$$

$$= \frac{g}{2} ((\partial_\mu W_{1\nu} - \partial_\nu W_{1\mu})(W_2^\mu W_3^\nu - W_3^\mu W_2^\nu)$$

$$+ (\partial_\mu W_{2\nu} - \partial_\nu W_{2\mu})(W_3^\mu W_1^\nu - W_1^\mu W_3^\nu)$$

$$+ (\partial_\mu W_{3\nu} - \partial_\nu W_{3\mu})(W_1^\mu W_2^\nu - W_2^\mu W_1^\nu)).$$

The third term has the correct form for a magnetic moment interaction with a magnetic field, after identifying $W_3 = \sin\theta_W A + \cos\theta_W Z$, and $g \sin\theta_W = e$. There are too many g's in this problem. The g in Equation (8.106) is not the g factor. There is another term hiding in the other two parts of Equation (8.106) that is identical to the third one, and can be extracted by parts integration. Multiplying the first two terms out gives:

$$\frac{g}{2} ((\partial_\mu W_{1\nu}) W_2^\mu W_3^\nu - (\partial_\mu W_{1\nu}) W_3^\mu W_2^\nu$$

$$- (\partial_\nu W_{1\mu}) W_2^\mu W_3^\nu + (\partial_\nu W_{1\mu}) W_3^\mu W_2^\nu$$

$$+ (\partial_\mu W_{2\nu}) W_3^\mu W_1^\nu - (\partial_\mu W_{2\nu}) W_1^\mu W_3^\nu$$

$$- (\partial_\nu W_{2\mu}) W_3^\mu W_1^\nu + (\partial_\nu W_{2\mu}) W_1^\mu W_3^\nu).$$

Integration by parts changes the signs, and switches the derivatives from one W to the product of two other W's. After expanding the derivatives of products, we have 16 terms, 12 not interesting for our purposes, and four that are identical to the third part of Equation (8.106). The expression of interest in the Lagrangian is therefore:

$$g(\partial_\mu W_{3\nu} - \partial_\nu W_{3\mu})(W_1^\mu W_2^\nu - W_2^\mu W_1^\nu)$$

$$= e(\partial_\mu A_\nu - \partial_\nu A_\mu)(W_1^\mu W_2^\nu - W_2^\mu W_1^\nu).$$

The factor of $1/2$ has gone away. We have dropped the Z term, and used $g \sin\theta_W = e$. The space–space parts of the field tensor are the components of the magnetic field:

$$\partial_i A_j - \partial_j A_i = \epsilon_{ijk} B_k;$$

giving

$$e\epsilon_{ijk} B_k (W_1^i W_2^j - W_2^i W_1^j).$$

Now use $W_1 = (W_+ + W_-)/\sqrt{2}$ and $W_2 = i(W_+ - W_-)/\sqrt{2}$ (Equation (7.84)) to write our magnetic moment term as a function of W_+ and W_-:

$$ie\epsilon_{ijk} B_k (W_-^i W_+^j - W_-^j W_+^i) = -2ie\epsilon_{ijk} B_k W_+^i W_-^j.$$

The W polarization states are:

$$\epsilon^{+1} = \frac{1}{\sqrt{2}}(0, 1, i, 0); \ \epsilon^{-1} = \frac{1}{2}(0, -1, i, 0) = -\epsilon^{+1*}; \ \vec{\epsilon}^+ \times \vec{\epsilon}^-$$

$$= \frac{1}{2}(i + i)\hat{z} = i\hat{z}.$$

Then $2e\vec{S} \cdot \vec{B}$ becomes $e\vec{S} \cdot \vec{B}/M_W$ when divided by the normalization $2M_W$; $g_W = 2$.

Chapter 9

Problem 9.1. The classical form of the Lenz vector defined for $L \neq 0$ is

$$\vec{M} = \frac{1}{m}\vec{p} \times \vec{L} - \frac{e^2 \vec{r}}{r}. \tag{9.129}$$

This is a constant vector in the plane of the orbit that is parallel to the semi-major axis of the ellipse, the line containing the foci. It represents the fact that the orientation of the ellipse remains fixed in space. For a $1/r^2$ central force there is no 'precession of the perihelion'. Show that $d\vec{M}/dt = 0$, if $d\vec{L}/dt = 0$ and $d\vec{p}/dt = -e^2\vec{r}/r^3$. The classical observables commute! The quantum mechanical observable is symmetrized to be Hermitian, because $(\vec{L} \times \vec{p})^\dagger = -\vec{p} \times \vec{L}$:

$$\vec{M} = \frac{1}{2m}(\vec{p} \times \vec{L} - \vec{L} \times \vec{p}) - \frac{e^2 \vec{r}}{r}$$

$$= \frac{1}{m}(\vec{p} \times \vec{L}) - \frac{i\hbar \vec{p}}{m} - \frac{e^2 \vec{r}}{r}. \tag{9.130}$$

$M^\dagger = M$. Some useful formulas are

$$\vec{L} \cdot \vec{M} = 0, \quad \vec{L} \times \vec{p} + \vec{p} \times \vec{L} = 2i\hbar\vec{p},$$

$$\vec{r} \cdot \vec{p} \times \vec{L} = \vec{L}^2, \text{ interchange dot and cross,}$$

and

$$[L_k, p_l] = i\hbar\epsilon_{klm}p_m, \quad \vec{p} \cdot \vec{L} = 0, \tag{9.131}$$

so that

$$(\vec{p} \times \vec{L})^2 = \sum \epsilon_{ijk}\epsilon_{ilm}p_j L_k p_l L_m = \vec{p}^{\,2}\vec{L}^2. \tag{9.132}$$

$\vec{M}$ is a constant of the motion, $[\vec{M}, H] = 0$, so the Lenz vector in quantum mechanics represents a degeneracy. Evaluation of the commutator is complicated because $\vec{p}$ and $\vec{r}$ do not commute. Use the definition of time dependence of an operator:

$$\frac{dM_x}{dt} = \frac{1}{i\hbar}[M_x, H] = 0 = \frac{1}{m}\left\{\frac{d(\vec{p} \times \vec{L})_x}{dt} - \frac{i\hbar dp_x}{dt}\right\} - e^2\frac{d}{dt}\frac{x}{r},$$

$$H = \frac{\vec{p}^{\,2}}{2m} - \frac{e^2}{r}. \tag{9.133}$$

Show that $1/m$ times the term in brackets is:

$$\frac{e^2}{mr^3}(r^2 p_x - x(\vec{r}\cdot\vec{p} - i\hbar)). \tag{9.134}$$

Then show that $e^2 d/dt(x/r)$ is the same, proving that the commutator vanishes.

Solution 9.1. Classical Lenz vector

$$\vec{M} = \frac{\vec{p} \times \vec{L}}{m} - \frac{e^2\vec{r}}{r}; \quad \frac{d\vec{M}}{dt} = \frac{1}{m}\frac{d\vec{p}}{dt}\times\vec{L} - e^2\left(\frac{d\vec{r}}{rdt} - \frac{\vec{r}\,dr}{r^2\,dt}\right); \quad \frac{d\vec{p}}{dt} = -\frac{e^2}{r^3}\vec{r};$$

and

$$\vec{r} \times (\vec{r} \times \vec{p}) = \vec{r}(\vec{r}\cdot\vec{p}) - \vec{p}r^2;$$

$$\frac{d\vec{M}}{dt} = -e^2\left(\frac{1}{r}\frac{d\vec{r}}{dt} - \frac{\vec{r}\,dr}{r^2\,dt} - \frac{1}{r}\frac{d\vec{r}}{dt} + \frac{\vec{r}\,dr}{r^2\,dt}\right) = 0.$$

If we adopt the suggested procedure, the quantum mechanical calculation is similar to the classical one. Return to Equation (9.133):

$$\frac{dM_x}{dt} = \frac{1}{i\hbar}[M_x, H] = 0 = \frac{1}{m}\left\{\frac{d(\vec{p} \times \vec{L})_x}{dt} - \frac{i\hbar dp_x}{dt}\right\} - e^2\frac{d}{dt}\frac{x}{r};$$

$$H = \frac{\vec{p}^{\,2}}{2m} - \frac{e^2}{r}.$$

The bracket term is:

$$\frac{1}{m}\left(\frac{dp_y}{dt}L_z - \frac{dp_z}{dt}L_y - i\hbar\frac{dp_x}{dt}\right); \quad \frac{dp_y}{dt} = \frac{1}{i\hbar}[p_y, H] = -e^2\frac{y}{r^3};$$

$$\frac{dp_z}{dt} = -e^2\frac{z}{r^3}; \quad \frac{dp_x}{dt} = -e^2\frac{x}{r^3}.$$

So

$$\frac{1}{m}\left(\frac{dp_y}{dt}L_z - \frac{dp_z}{dt}L_y - i\hbar\frac{dp_x}{dt}\right) = -\frac{e^2}{mr^3}(yL_z - zL_y - i\hbar x)$$

$$= -\frac{e^2}{mr^3}(-r^2 p_x + x(\vec{r}\cdot\vec{p}) - i\hbar x).$$

This is the same as Equation (9.134). Notice that in these manipulations non-commuting variables have not been interchanged, and the derivative operators are on the right-hand side. We now have to show that the last term cancels the bracket, and this is more complicated, because of $\vec{p}^{\,2}$ in the Hamiltonian.

$$-e^2\frac{d}{dt}\frac{x}{r} = -\frac{e^2}{2mi\hbar}\left[\frac{x}{r}, p_x^2 + p_y^2 + p_z^2\right]$$

$$= -\frac{e^2}{2mi\hbar}\left\{p_x[x/r, p_x] + [x/r, p_x]p_x + p_y[x/r, p_y]\right.$$
$$\left. + [x/r, p_y]p_y + p_z[x/r, p_z] + [x/r, p_z]p_z\right\}.$$

Evaluating the commutators is a chore, but you aren't finished, because you have to move p_x etc through from left to right. However, it works, so soldier on.

$$[x/r, p_x] = -\frac{i\hbar x^2}{r^3} + \frac{i\hbar}{r}; \quad [x/r, p_y] = -\frac{i\hbar xy}{r^3}; \quad [x/r, p_z] = -\frac{i\hbar xz}{r^3}.$$

Now move the operators through the commutators:

$$p_y[x/r, p_y] + [x/r, p_y]p_y = -\frac{\hbar^2 x}{r^3} + \frac{3\hbar^2 xy^2}{r^5} - \frac{2i\hbar xy}{r^3}p_y;$$

$$p_z[x/r, p_z] + [x/r, p_z]p_z = -\frac{\hbar^2 x}{r^3} + \frac{3\hbar^2 xz^2}{r^5} - \frac{2i\hbar xz}{r^3}p_z;$$

and finally

$$p_x[x/r, p_x] + [x/r, p_x]p_x = -\frac{3\hbar^2 x}{r^3} + \frac{3\hbar^2 x^3}{r^5} - \frac{2i\hbar x^2}{r^3}p_x + \frac{2i\hbar}{r}p_x.$$

The operators are now on the right-hand side of all of the coordinates. Adding the terms, and multiplying by $-e^2/(2mi\hbar)$ cancels Equation (9.134), proving $\left[\vec{M}, H\right] = 0$.

Problem 9.2. Work out the quantum numbers in spherical coordinates for the third excited state of the three dimensional isotropic harmonic oscillator: $n_x + n_y + n_z = 3$. Match degeneracy, parity, and E_n. The radial wave function for a three-dimensional harmonic oscillator may be written, following Problem 10.10 in Merzbacher:

$$\psi(r, \theta, \phi) = r^l e^{-\alpha^2 r^2/2} f(r) Y_l^{m_l}(\theta, \phi),$$

$$\alpha^4 = \frac{mk}{\hbar^2}, \quad \omega^2 = \frac{k}{m}. \tag{9.135}$$

The radial wave equation is

$$\left\{ \frac{d^2}{dr^2} + \frac{2d}{rdr} - \left(\frac{mkr^2}{\hbar^2} + \frac{l(l+1)}{r^2} - \frac{2mE}{\hbar^2} \right) \right\} R(r) = 0. \tag{9.136}$$

Substitute $R(r)$ from Equation (9.135) to obtain:

$$f''(r) + \left(\frac{2(l+1)}{r} - 2\alpha^2 r \right) f'(r)$$

$$+ 4\alpha^2 \left(\frac{E}{2\hbar\omega} - \frac{(2l+3)}{4} \right) f(r) = 0. \tag{9.137}$$

The Gaussian asymptotic form takes out the harmonic oscillator potential, and r^l takes out the centrifugal potential. Hydrogen atom wave functions resemble Equation (9.135), except that a damped exponential replaces the Gaussian. This is a clue that a change of variables $r \to r^2$ is needed to relate the two problems. This is what is meant in the text by 'judicious choice of variables'. Associated Laguerre polynomials $L_{q-p}^p(z)$ satisfy the differential equation:

$$z\frac{d^2w}{dz^2} + (p+1-z)\frac{dw}{dz} + (q-p)w = 0. \tag{9.138}$$

Replace $z \to \alpha^2 r^2$ and obtain the form:

$$\frac{d^2w}{dr^2} + \left(\frac{2p+1}{r} - 2\alpha^2 r \right)\frac{dw}{dr} + 4\alpha^2(q-p)w = 0. \tag{9.139}$$

Comparison of Equations (9.137) and (9.139) shows that $p = l + 1/2$ and $q - p = n - 1$, where $E_n = \hbar\omega(2n + l - 1/2)$. The indices q and p are half integers, but $q - p$ is an integer. The solutions are generalizations of associated Laguerre polynomials for half integer p. This is the connection between hydrogen and the harmonic oscillator.

Solution 9.2. Three-dimensional symmetric harmonic oscillator. In Cartesian coordinates:

$$E_{n_x,n_y,n_z} = \hbar\omega(n_x + n_y + n_z + 3/2);$$

and in spherical coordinates:

$$E_{n,l} = \hbar\omega(2n + l - 1/2); \quad 2n + l = n_x + n_y + n_z + 2; \quad n_x + n_y + n_z$$
$$= 3 \rightarrow 2n + l = 5.$$

Parity odd.

$$n_x + n_y + n_z = 3; \ 1 + 1 + 1 \text{ one state}; \ 3 + 0 + 0 \text{ three states};$$
$$2 + 1 + 0 \text{ six states}; \text{ degeneracy} = 10.$$

So $l = 1$ or $l = 3$. To add up to 5, $n = 2, l = 1$ or $n = 1, l = 3$.

$$\psi(r,\theta,\phi) = r^l e^{-\alpha^2 r^2/2} f(r) Y_l^{m_l}(\theta,\phi); \quad \alpha^4 = \frac{mk}{\hbar^2}; \quad \omega^2 = \frac{k}{m}.$$

Substitute $R(r) = r^l e^{-\alpha^2 r^2/2} f(r)$ into Equation (9.136):

$$\left\{ \frac{d^2}{dr^2} + \frac{2d}{rdr} - \left(\frac{mkr^2}{\hbar^2} + \frac{l(l+1)}{r^2} - \frac{2mE}{\hbar^2} \right) \right\} R(r) = 0.$$

$$\frac{d^2 R}{dr^2} = r^l e^{-\alpha^2 r^2/2} \left\{ f''(r) + \left(\frac{2l}{r} - 2\alpha^2 r \right) f'(r) \right.$$
$$\left. + \left(\frac{l(l-1)}{r^2} + \alpha^4 r^2 - \alpha^2(2l+1) \right) f(r) \right\}.$$

$$\frac{2}{r}\frac{dR}{dr} = r^l e^{-\alpha^2 r^2/2} \left\{ \frac{2}{r} f'(r) + \left(\frac{2l}{r^2} - 2\alpha^2 \right) f(r) \right\}.$$

Combining these two terms with

$$-r^l e^{-\alpha^2 r^2/2} \left\{ \alpha^4 r^2 + \frac{l(l+1)}{r^2} - \frac{2mE}{\hbar^2} \right\} f(r)$$

gives

$$f''(r) + \left(\frac{2(l+1)}{r} - 2\alpha^2 r\right) f'(r) + 4\alpha^2 \left(\frac{E}{2\hbar\omega} - \frac{(2l+3)}{4}\right) f(r) = 0.$$

Then substituting $E_{n,l} = \hbar\omega(2n + l - 1/2)$ gives:

$$f''(r) + \left(\frac{2(l+1)}{r} - 2\alpha^2 r\right) f'(r) + 4\alpha^2 (n-1) f(r) = 0.$$

Now take Equation (9.138) and substitute $z \to \alpha^2 r^2$:

$$\frac{dw}{dz} = \frac{1}{2\alpha^2 r}\frac{dw}{dr}; \quad \frac{d^2 w}{dz^2} = \frac{1}{4\alpha^4 r^2}\left(\frac{d^2 w}{dr^2} - \frac{1}{r}\frac{dw}{dr}\right).$$

Substitution into the equation for the associated Laguerre polynomials gives:

$$z\frac{d^2 w}{dz^2} + (p+1-z)\frac{dw}{dz} + (q-p)w = \frac{1}{4\alpha^2}\left(\frac{d^2 w}{dr^2} - \frac{1}{r}\frac{dw}{dr}\right)$$

$$+ \frac{p+1-\alpha^2 r^2}{2\alpha^2 r}\frac{dw}{dr} + (q-p)w = 0.$$

Multiply by $4\alpha^2$ and combine terms to get:

$$\frac{d^2 w}{dr^2} + \left(\frac{2p+1}{r} - 2\alpha^2 r\right)\frac{dw}{dr} + 4\alpha^2 (q-p)w = 0.$$

Comparison shows that $p = l + 1/2$ and $q - p = n - 1$.

Problem 9.3. Contribution of the annihilation diagram $e^- e^+ \to \gamma \to e^- e^+$ to the positronium ground state. First we need a relation between the relativistic T matrix element calculated by the Feynman rules, and a non-relativistic potential $V(r)$. The link is made via the Born approximation for scattering. Standard quantum mechanics texts (Merzbacher, Chapter 11, Section 4) give the scattering cross-section in terms of the potential and the momentum transfer in the center of mass as

$$\frac{d\sigma}{d\Omega} = \frac{(m_r c^2)^2}{4\pi^2 (\hbar c)^4} \left| \int e^{i(\vec{k}-\vec{k}')\cdot\vec{r}} V(\vec{r}) d^3 r \right|^2. \qquad (9.140)$$

For Bhabha scattering the reduced mass $m_r = m_e/2$. Compare this to Equation (4.69) in cgs units:

$$\frac{d\sigma}{d\Omega} = \frac{\hbar^2 c^2}{64\pi^2 s} |\langle f | T | i \rangle|^2. \qquad (9.141)$$

In the non-relativistic limit for Bhabha scattering $s = (2m_e c^2)^2$. Comparing these two equations gives:

$$\int e^{i(\vec{k}-\vec{k}')\cdot\vec{r}} V(\vec{r}) d^3 r = \frac{(\hbar c)^3}{4(m_e c^2)^2} \langle f | T | i \rangle. \qquad (9.142)$$

The potential is given by the Fourier transform:

$$V(r) = \frac{(\hbar c)^3}{4(m_e c^2)^2} \frac{1}{(2\pi)^3} \int d^3 q \, e^{-i\vec{q}\cdot\vec{r}} \langle f | T | i \rangle, \quad \vec{q} = \vec{k} - \vec{k}'. \qquad (9.143)$$

Here, it is understood that the T matrix element is evaluated in the non-relativistic limit. The sign of the T matrix element now becomes important, because the potential can be either attractive or repulsive. The Feynman rules keep track of whether the force is attractive $T < 0$, or repulsive $T > 0$. This can be demonstrated in the non-relativistic approximation by relating T to H (see Equation (4.39) in Sakurai 'Advanced Quantum Mechanics'). Cross-sections in first-order covariant perturbation theory, like the Born approximation, are independent of the sign of the charge, but this is not true for higher-order corrections, and the cross-sections for e^- and e^+ on high Z nuclei are quite different.

Evaluate T in the non-relativistic limit for Bhabha scattering by single photon exchange:

$$\langle f | T | i \rangle = -\frac{e^2}{q^2} \bar{u}(p_3) \gamma^\mu u(p_1) \bar{v}(p_2) \gamma_\mu v(p_4). \qquad (9.144)$$

Obtain the non-relativistic limit averaged over initial spins and summed over final spins:

$$\frac{1}{4}\sum_{\text{spins}} \langle f|T|i\rangle = -\frac{4e^2 m_e^2}{q^2} = -\frac{16\pi (m_e c^2)^2 \alpha}{\hbar^2 c^2 q^2}, \tag{9.145}$$

where we have used the fine structure constant to help convert from natural units to cgs units, and $\vec{q}$ is now a wave number, with dimensions $1/l$. The T matrix element is still dimensionless, and $\alpha = e^2/(\hbar c) = 1/137$. Substitute this in Equation (9.143), and use $\int_0^\infty dx(\sin x)/x = \pi/2$ to obtain:

$$V(r) = -\frac{e^2}{r}. \tag{9.146}$$

So this works. Now apply the same analysis to the annihilation diagram:

$$\langle f|T|i\rangle = -\frac{e^2}{q^2}\bar{v}(p_2)\gamma^\mu u(p_1)\bar{u}(p_3)\gamma_\mu v(p_4). \tag{9.147}$$

There is a sign difference that comes from the matching of e^- and e^+ spinors, and replacing γ^0 with $\vec{\gamma}$. The other important difference is that the propagator is now a constant: $q^2 = 4m_e^2$, and the effective potential is a $\delta(\vec{r})$ function. Show that only triplet states contribute, and obtain the result for the expectation value of the annihilation potential for the positronium ground state wave function:

$$\langle V_{ann}\rangle = \frac{(\hbar c)^3 \pi \alpha}{(m_e c^2)^2}|\psi_{n=1,l=0}(0)|^2 = \frac{m_e c^2 \alpha^4}{4}. \tag{9.148}$$

Solution 9.3. Annihilation diagram contribution to the positronium hyperfine structure. Equation (9.143) reads:

$$V(r) = \frac{(\hbar c)^3}{4(m_e c^2)^2}\frac{1}{(2\pi)^3}\int d^3q\, e^{-i\vec{q}\cdot\vec{r}}\langle f|T|i\rangle\,;\ \vec{q}=\vec{k}-\vec{k}'.$$

The T matrix element for $e^- + e^+ \to \gamma \to e^- + e^+$ is:

$$\langle f|T|i\rangle = -\frac{e^2}{q^2}(\bar{v}^{s_1'}(p_2)\gamma^\mu u^{s_1}(p_1))(\bar{u}^{s_2}(p_3)\gamma_\mu v^{s_2'}(p_4)).$$

All of the four momenta are $p = (m, 0, 0, 0)$, and $q^2 = 4m^2$, where m is the electron mass. So the spinors are:

$$u^{(1)}(m) = \sqrt{2m} \begin{pmatrix} 1 \\ 0 \\ 0 \\ 0 \end{pmatrix}; \quad u^{(2)}(m) = \sqrt{2m} \begin{pmatrix} 0 \\ 1 \\ 0 \\ 0 \end{pmatrix};$$

$$v^{(1)}(m) = \sqrt{2m} \begin{pmatrix} 0 \\ 0 \\ 0 \\ 1 \end{pmatrix}; \quad v^{(2)}(m) = \sqrt{2m} \begin{pmatrix} 0 \\ 0 \\ 1 \\ 0 \end{pmatrix};$$

and the adjoint spinors are:

$$\bar{u}^{(1)} = \sqrt{2m} \begin{pmatrix} 1 & 0 & 0 & 0 \end{pmatrix}; \quad \bar{u}^{(2)} = \sqrt{2m} \begin{pmatrix} 0 & 1 & 0 & 0 \end{pmatrix};$$

$$\bar{v}^{(1)} = \sqrt{2m} \begin{pmatrix} 0 & 0 & 0 & -1 \end{pmatrix}; \quad \bar{v}^{(2)} = \sqrt{2m} \begin{pmatrix} 0 & 0 & -1 & 0 \end{pmatrix}.$$

γ^0 changes the signs of the lower components of v. Upper and lower components have to be mixed in order to get a non-zero result, so γ^0 plays no role in the current, and we may write:

$$\langle f | T | i \rangle = \frac{e^2}{4m^2} (\bar{v}^{s_1'}(m) \vec{\gamma} u^{s_1}(m)) \cdot (\bar{u}^{s_2}(m) \vec{\gamma} v^{s_2'}(m)).$$

The overall sign is changed by the metric tensor. All of the spin combinations have to be considered. The matrices are:

$$\gamma^x = \begin{pmatrix} 0 & 0 & 0 & 1 \\ 0 & 0 & 1 & 0 \\ 0 & -1 & 0 & 0 \\ -1 & 0 & 0 & 0 \end{pmatrix}; \quad \gamma^y = \begin{pmatrix} 0 & 0 & 0 & -i \\ 0 & 0 & i & 0 \\ 0 & i & 0 & 0 \\ -i & 0 & 0 & 0 \end{pmatrix}$$

$$\gamma^z = \begin{pmatrix} 0 & 0 & 1 & 0 \\ 0 & 0 & 0 & -1 \\ -1 & 0 & 0 & 0 \\ 0 & 1 & 0 & 0 \end{pmatrix}.$$

Multiplying parts together gives the following initial state matrix elements:

$$\bar{v}^{(1)} \gamma^x u^{(1)} = 2m; \quad \bar{v}^{(1)} \gamma^y u^{(1)} = 2mi; \quad \bar{v}^{(1)} \gamma^z u^{(1)} = 0;$$

$$\bar{v}^{(2)} \gamma^x u^{(1)} = \bar{v}^{(2)} \gamma^y u^{(1)} = 0; \quad \bar{v}^{(2)} \gamma^z u^{(1)} = 2m;$$

$$\bar{v}^{(2)}\gamma^x u^{(2)} = 2m; \ \bar{v}^{(2)}\gamma^y u^{(2)} = -2mi; \ \bar{v}^{(2)}\gamma^z u^{(2)} = 0;$$
$$\bar{v}^{(1)}\gamma^x u^{(2)} = \bar{v}^{(1)}\gamma^y u^{(2)} = 0; \ \bar{v}^{(1)}\gamma^z u^{(2)} = -2m;$$

and the final state

$$\bar{u}^{(1)}\gamma^x v^{(1)} = 2m; \ \bar{u}^{(1)}\gamma^y v^{(1)} = -2mi; \ \bar{u}^{(1)}\gamma^z v^{(1)} = 0;$$
$$\bar{u}^{(2)}\gamma^x v^{(1)} = \bar{u}^{(2)}\gamma^y v^{(1)} = 0; \ \bar{u}^{(2)}\gamma^z v^{(1)} = -2m;$$
$$\bar{u}^{(2)}\gamma^x v^{(2)} = 2m; \ \bar{u}^{(2)}\gamma^y v^{(2)} = 2mi; \ \bar{u}^{(2)}\gamma^z v^{(2)} = 0;$$
$$\bar{u}^{(1)}\gamma^x v^{(2)} = \bar{u}^{(1)}\gamma^y v^{(2)} = 0; \ \bar{u}^{(1)}\gamma^z v^{(2)} = 2m.$$

The total spin magnetic quantum number must be the same in the initial and final states. Then we have:

$$\sum_{\text{spins}} (\bar{v}\gamma^x u)(\bar{u}\gamma^x v) = 8m^2; \ \sum_{\text{spins}} (\bar{v}\gamma^y u)(\bar{u}\gamma^y v) = 8m^2;$$

$$\sum_{\text{spins}} (\bar{v}\gamma^z u)(\bar{u}\gamma^z v) = 0;$$

and the T matrix element divided by two to average over initial state spins ($m_s = \pm 1$) is:

$$\frac{1}{2} \sum_{\text{spins}} \langle f | T | i \rangle = 8\pi\alpha.$$

Only the triplet state contributes, which has to be true for the single intermediate photon. Substitution into Equation (9.143), and calculating the expectation value of $V(r)$ using the hydrogen-like wave function gives Equation (9.148).

$$V(r) = \frac{(\hbar c)^3}{4(m_e c^2)^2} \frac{1}{(2\pi)^3} \int d^3 q e^{-i\vec{q}\cdot\vec{r}} \langle f | T | i \rangle = \frac{2\pi\alpha(\hbar c)^3}{(m_e c^2)^2} \delta(\vec{r}).$$

Then

$$<V>_{n=1, l=0} = \frac{2\pi\alpha(\hbar c)^3}{(m_e c^2)^2} \frac{1}{\pi(2a_0)^3}; \ \text{where}$$

$$a_0 = \frac{\hbar c}{\alpha m_e c^2} \ \text{is the electron Bohr radius.}$$

Cancelling terms gives the desired result:

$$< V > = \frac{\alpha^4 m_e c^2}{4} = 3.6 \times 10^{-4} \text{ eV};$$

when added to the "usual" hyperfine splitting formula (see the next problem) this gives the triplet-singlet splitting of the positronium ground state of 8.4×10^{-4} eV.

Problem 9.4. This problem builds on the previous one to calculate the singlet–triplet splitting of the positronium ground state. The annihilation term Equation (9.148) gets added to the usual magnetic moment interaction term. The vector potential from a classical magnetic moment $\vec{m}$ at the origin is given in cgs units by the formula:

$$\vec{A} = -\vec{m} \times \vec{\nabla}\left(\frac{1}{r}\right), \quad \text{so } \vec{B} = -\vec{\nabla} \times \vec{m} \times \vec{\nabla}\left(\frac{1}{r}\right). \qquad (9.149)$$

The classical magnetic moment is replaced by the positron magnetic moment operator $\vec{\mu}_{pos} = e\hbar/(2m_e c)\vec{\sigma}_2$ in the quantum mechanical Hamiltonian. For the positronium hyperfine structure we write:

$$H_{int} = -\vec{\mu}_{ele} \cdot \vec{B} = -\left(\frac{e\hbar}{2m_e c}\right)^2 \vec{\sigma}_1 \cdot (\vec{\nabla} \times \vec{\sigma}_2 \times \vec{\nabla})\left(\frac{1}{r}\right). \qquad (9.150)$$

The two magnetic moments are equal in magnitude and opposite in sign. The triple cross product can be rewritten using the usual formula $\vec{a} \times (\vec{b} \times \vec{c}) = \vec{b}(\vec{a} \cdot \vec{c}) - \vec{c}(\vec{a} \cdot \vec{b})$, which is the same as $\epsilon_{ijk}\epsilon_{ilm} = \delta_{jl}\delta_{km} - \delta_{jm}\delta_{kl}$. Thus,

$$H_{int} = -\left(\frac{e\hbar}{2m_e c}\right)^2 (\vec{\sigma}_1 \cdot \vec{\sigma}_2 \vec{\nabla}^2 - (\sigma_1 \cdot \vec{\nabla})(\sigma_2 \cdot \vec{\nabla}))\frac{1}{r}, \qquad (9.151)$$

and the second term is

$$-(\vec{\sigma}_1 \cdot \vec{\nabla})(\vec{\sigma}_2 \cdot \vec{\nabla})\left(\frac{1}{r}\right) = -\left(\sigma_{1x}\frac{\partial}{\partial x} + \sigma_{1y}\frac{\partial}{\partial y} + \sigma_{1z}\frac{\partial}{\partial z}\right)$$

$$\times \left(\sigma_{2x}\frac{\partial}{\partial x} + \sigma_{2y}\frac{\partial}{\partial y} + \sigma_{2z}\frac{\partial}{\partial z}\right)\left(\frac{1}{r}\right). \qquad (9.152)$$

Equation (9.152) has to be evaluated carefully. 'Diagonal' terms like $(\sigma_{1x}\sigma_{2x}\partial^2/\partial x^2)(1/r) = (1/3)\sigma_{1x}\sigma_{2x}\vec{\nabla}^2(1/r)$, so the sum of diagonal terms subtracts $1/3$ from the first term. The space parts of off diagonal terms like $\sigma_{1x}\sigma_{2y}\partial^2/\partial x\partial y(1/r)$ transform like the Y_2^m spherical harmonics. Since we are interested in the ground state of positronium that has $l = 0$, we subtract $(\vec{\sigma}_1 \cdot \vec{\sigma}_2\vec{\nabla}^2)/3$ from the first term, and then ignore what remains of the second part.

$$H_{\text{int}} = -\frac{2}{3}\left(\frac{e\hbar}{2m_e c}\right)^2 (\vec{\sigma}_1 \cdot \vec{\sigma}_2)\vec{\nabla}^2\left(\frac{1}{r}\right). \qquad (9.153)$$

$\vec{\nabla}^2(1/r) = -4\pi\delta(\vec{r})$ is the singularity referred to in the text. Evaluate the singlet–triplet splitting using the Hamiltonian in Equation (9.153), and combine it with the annihilation diagram Equation (9.148) to obtain the final result:

$$\Delta E_{hfs} = E(n = 1, l = 0, \text{triplet}) - E(n = 1, l = 0, \text{singlet})$$

$$= \left(\frac{1}{3} + \frac{1}{4}\right)\alpha^4 m_e c^2. \qquad (9.154)$$

Give the number in eV and in MHz. This number can be accurately measured. Early experiments have been described by Deutsch [16]. About 30% of positrons stopping in a gas form positronium. The rest annihilate in flight. The triplet/singlet ratio is 3/1. Triplet positronium decays into three γ rays that give a continuous energy spectrum, while the singlet state gives a line spectrum at 511 keV. Thus, microwaves of the correct frequency will cause the triplet to singlet transition, and a decrease in the continuous γ spectrum. Deutsch's original measurements were made in a magnetic field where the Zeeman splitting between $J = 1, m = \pm 1$ and $J = 1, m = 0$ was measured — a smaller energy difference than Equation (9.154). The reason for application of the Zeeman effect was that at that time microwave sources in the 1 mm wavelength region were not available, but in the 2 cm range they were.

Solution 9.4. We need the expectation value of Equation (9.153):

$$H_{int} = -\frac{2}{3}\left(\frac{e\hbar}{2m_e c}\right)^2 (\vec{\sigma}_1 \cdot \vec{\sigma}_2)\vec{\nabla}^2\left(\frac{1}{r}\right).$$

Using the identity $\nabla^2(1/r) = -4\pi\delta(\vec{r})$, and the positronium ground state wave function gives:

$$< H_{\text{int}} >= \frac{(e\hbar)^2}{3(m_e c)^2} \frac{\alpha^3 (m_e c^2)^3}{(\hbar c)^3} (s(s+1) - 3/2)/2.$$

The triplet-singlet splitting is just the constant, which after cancellations becomes:

$$\Delta H = \frac{\alpha^4 m_e c^2}{3} = 4.8 \times 10^{-4} \text{eV}.$$

This is to be added to the result of Problem 9.3.

Problem 9.5. Zeeman effect in positronium. Positronium has the curious feature that it has no linear Zeeman effect. That means that it has no permanent magnetic dipole moment. Since e^- and e^+ chase each other around the orbit, there is no current associated with the angular momentum $\vec{L}$, and no orbital magnetic moment. There is no spin magnetic moment either. Obviously for $S = 1$ there is no moment, because if $m_1 = +1/2$ and $m_2 = +1/2$ the magnetic moments cancel, and there is no net effect for $m_s = 0$. The effect is for $m_1 = +1/2$ and $m_2 = -1/2$ or vice versa, decoupling the total S eigenstates. The ground state interaction term in a magnetic field is

$$H_B = -(\vec{\mu}_+ + \vec{\mu}_-) \cdot \vec{B} = -\frac{|e|}{mc}(\vec{S}_+ - \vec{S}_-) \cdot \vec{B}. \tag{9.155}$$

Let m_1 be the electron and m_2 the positron. Then it is straightforward to calculate the matrix elements:

$$\langle m_1 = +1/2, \ m_2 = -1/2| \ H_B \ |m_1 = +1/2, \ m_2 = -1/2\rangle$$
$$= \frac{|e|\hbar}{mc} B_z = \Delta_B,$$

and

$$\langle m_1 = -1/2, \ m_2 = +1/2| \ H_B \ |m_1 = -1/2, \ m_2 = +1/2\rangle$$
$$= -\Delta_B. \tag{9.156}$$

Diagonalize the matrix in m_1, m_2 space combining Equations (9.155) and (9.156). It is a 2×2 matrix, because $+1/2, +1/2$ and $-1/2, -1/2$ are decoupled. Obtain the eigenvalues:

$$\lambda_\pm = (E_1 + E_0)/2 \pm \sqrt{(E_1 - E_0)^2/4 + \Delta_B^2}, \qquad (9.157)$$

where $E_1 - E_0 = \Delta E_{hfs}$, and E_1 is the $S = 1$ energy level.

Expand the square root for small $\Delta_B/\Delta E_{hfs}$, and show that the $S = 1, m_s = 0$ level and the $S = 0, m_s = 0$ level spread apart as a quadratic function of the magnetic field. The original measurement of ΔE_{hfs} was done using the formula for λ_+, and exploiting the fact that $S = 1$ $m_s = 0$ decoupled into ↑↓ and ↓↑ states causes $S = 0$ to appear, with the resulting two γ decay. The rf frequency was fixed, and the magnetic field varied to observe a decrease in the 3γ decays, that have a continuous energy spectrum (2γ decays are all at 511 keV). The frequency was then inversely proportional to ΔE_{hfs}, and the numerator Δ_B was known.

Solution 9.5. Quadratic Zeemann effect in positronium. As explained in the problem, you have to break apart the $S = 1, m_s = 0$ and $S = 0, m_s = 0$ states in order to couple the magnetic moments to an external magnetic field. The Hamiltonian is Equation (9.155):

$$H_B = -(\vec{\mu}_+ + \vec{\mu}_-) \cdot \vec{B} = -\frac{|e|}{mc}(\vec{S}_+ - \vec{S}_-) \cdot \vec{B}.$$

The triplet and singlet states are split as we have seen by 8.4×10^{-4} eV. Call the energy levels E_1 and E_0. Use the Clebsch–Gordan coefficients to write these levels in the (m_1, m_2) representation, where m_1 is the magnetic quantum number of the electron. This gives the following matrix (without H_B):

$$\begin{pmatrix} & ++ & -- & +- & -+ \\ ++ & E_1 & 0 & 0 & 0 \\ -- & 0 & E_1 & 0 & 0 \\ +- & 0 & 0 & (E_1 + E_0)/2 & (E_1 - E_0)/2 \\ -+ & 0 & 0 & (E_1 - E_0)/2 & (E_1 + E_0)/2 \end{pmatrix}.$$

The matrix elements of H_B in the (m_1, m_2) representation populate the diagonal elements of the lower 2x2 sub matrix, leaving the $m_s =$

± 1 terms alone. For $\vec{B} = B\hat{z}$ the H_B matrix is diagonal:

$$\langle m_1, m_2 | H_B | m_1, m_2 \rangle = -\frac{|e|\hbar B}{mc}(m_2 - m_1); \text{ define } \Delta = \frac{|e|\hbar B}{mc};$$

Then the diagonalization determinant is:

$$\begin{bmatrix} (E_1 + E_0)/2 + \Delta - \lambda & (E_1 - E_0)/2 \\ (E_1 - E_0)/2 & (E_1 + E_0)/2 - \Delta - \lambda \end{bmatrix} = 0.$$

Giving the quadratic equation:

$$\lambda^2 - \lambda(E_1 + E_0) + E_1 E_0 - \Delta^2 = 0;$$

$$\lambda_\pm = \frac{E_1 + E_0}{2} \pm \frac{\sqrt{(E_1 - E_0)^2 + 4\Delta^2}}{2}.$$

The hyperfine splitting is $E_1 - E_0$. If $\Delta/(E_1 - E_0) << 1$, then:

$$\lambda_\pm = \frac{E_1 + E_0}{2} \pm \frac{E_1 - E_0}{2}\left(1 + \frac{2\Delta^2}{(E_1 - E_0)^2} + \cdots\right);$$

or

$$\lambda_+ = E_1 + \frac{\Delta^2}{E_1 - E_0}; \quad \lambda_- = E_0 - \frac{\Delta^2}{E_1 - E_0}.$$

The splitting is quadratic in the magnetic field. One line curves upwards from E_1, and the other curves downwards from E_0. A sketch of the splitting is shown in Figure 1.

Problem 9.6. Muonium precession in an external magnetic field. Assume the muon is 100% polarized along its direction of motion. When the muon stops in material and captures an electron, half of the electron spins will be parallel to the muon spin, and half antiparallel. The antiparallel spins will form atoms with $m_s = 0$, so half of the muons will be depolarized. The other half will form muonium with $S = 1$, and $\langle S_y \rangle = +\hbar$ at $t = 0$. Show that the spin motion

Zeeman Effect in Positronium Ground State

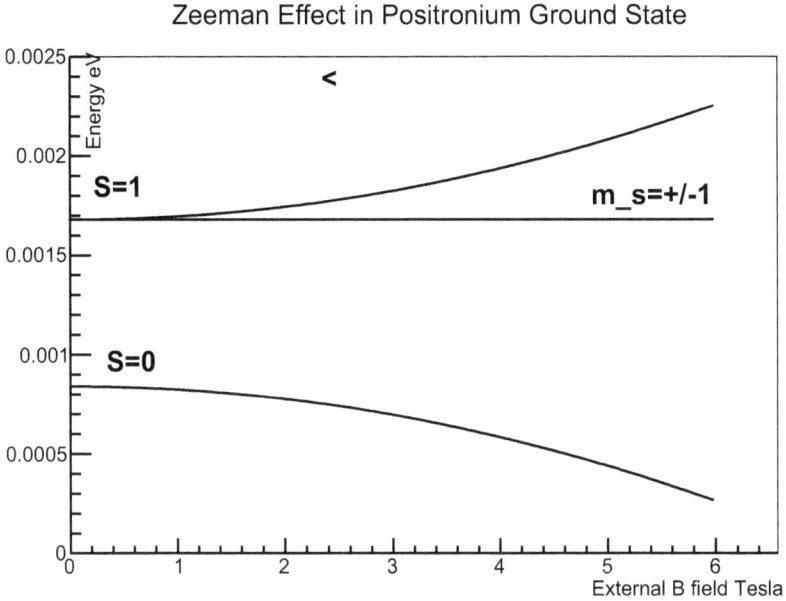

Figure 1: Quadratic Zeeman effect in the positronium ground state. Magnetic field in Tesla is on the x-axis. At B=0, the $S = 0$ and $S = 1$ levels are split by the hyperfine splitting of 8.4×10^{-4} eV. The energy of the singlet state is arbitrary. The triplet states with $m_s = \pm 1$ are not affected by the field. Triplet and singlet states with $m_s = 0$ are decoupled into states $|\uparrow\downarrow\rangle$ and $|\downarrow\uparrow\rangle$, that have twice the electron's magnetic moment

Hamiltonian is:

$$H = -\vec{\mu} \cdot \vec{B} = -\frac{eB}{m_e c}\left(-S_z + \left(1 + \frac{m_e}{m_\mu}\right)S_{2z}\right), \qquad (9.158)$$

where $\vec{S} = \vec{S}_1 + \vec{S}_2$, and S_2 is the muon. Using the commutation relations, show that

$$\frac{dS_y}{dt} = \frac{eB}{2m_e c}\left(1 - \frac{m_e}{m_\mu}\right)S_x, \quad \text{and}$$

$$\frac{dS_x}{dt} = -\frac{eB}{2m_e c}\left(1 - \frac{m_e}{m_\mu}\right)S_y. \qquad (9.159)$$

Hence the precession frequency is $\omega = (eB)/(2m_e c)(1 - m_e/m_\mu)$, about 100 times the muon frequency. The magnetic moment is essentially that of the electron, but the angular momentum vector is twice as large.

Solution 9.6. Muonium precession in an external magnetic field. To obtain Equation (9.158), write:

$$\vec{\mu} = \vec{\mu}_e + \vec{\mu}_\mu = -\frac{e}{m_e c}\vec{S}_e + \frac{e}{m_\mu c}\vec{S}_\mu.$$

Then with the electron being particle 1 and the muon particle 2, $\vec{B} = B\hat{z}$, and $S_{1z} = S_z - S_{2z}$ we have:

$$H = -\vec{\mu} \cdot \vec{B} = -\frac{eB}{m_e c}\left(-S_z + \left(1 + \frac{m_e}{m_\mu}\right)S_{2z}\right);$$

Now use the definition:

$$\frac{dS_y}{dt} = \frac{1}{i\hbar}[S_y, H]; \quad [S_y, S_z] = i\hbar S_x;$$

$$[S_y, S_{2z}] = [S_y, S_z] - [S_y, S_{1z}]; \quad \text{and}$$

$$[S_y, S_{1z}] = [S_y, S_{2z}].$$

So

$$[S_y, S_{2z}] = \frac{1}{2}[S_y, S_z]; \quad \text{giving} \quad \frac{dS_y}{dt} = \frac{eB}{2m_e c}\left(1 - \frac{m_e}{m_\mu}\right)S_x, \quad \text{and}$$

$$\frac{dS_x}{dt} = -\frac{eB}{2m_e c}\left(1 - \frac{m_e}{m_\mu}\right)S_y.$$

Ignoring $m_e/m_\mu = 0.005$, the muonium precession frequency is half that of the free electron. Muonium has the electron's magnetic moment, but twice the electron's spin, hence the factor of two.

Problem 9.7. Rate for muonium annihilation: $\mu^+ e^- \rightarrow \bar{\nu}_\mu \nu_e$. This is similar to $e^- e^+ \rightarrow \gamma\gamma$, only using the weak rather than the electromagnetic interaction. There is only one W t channel exchange diagram, because the Z does not couple the muon to an electron, so the problem is simpler than positronium. The momenta are: electron $p_1 = (m_e, 0, 0, 0)$; muon $p_2 = (m_\mu, 0, 0, 0)$; ν_e $p_3 = (m_\mu/2, 0, 0, m_\mu/2)$; and $\bar{\nu}_\mu$ $p_4 = (m_\mu/2, 0, 0, -m_\mu/2)$. The total energy is really $\sqrt{s} = m_e + m_\mu$, but we dropped the electron mass, and we chose the z axis for the two final state neutrino momenta. Check that the Feynman rules give the T matrix element:

$$-i\,\langle \nu_e \cdot \bar{\nu}_\mu | \, T \, | e^- \mu^+ \rangle$$

$$= \bar{u}(p_3)\frac{-ig}{\sqrt{2}}\gamma^\mu \frac{1-\gamma^5}{2} u(p_1)\frac{-i(g_{\mu\nu} + q_\mu q_\nu / M_W^2)}{q^2 - M_W^2}$$

$$\times \bar{v}(p_2)\frac{-ig}{\sqrt{2}}\gamma^\nu \frac{1-\gamma^5}{2} v(p_4). \tag{9.160}$$

$|q^2| = |(p_1 - p_3)^2| \ll M_W^2$, so replace the W propagator with $ig_{\mu\nu}/M_W^2$, and then use $G_F/\sqrt{2} = g^2/(8M_W^2)$ to introduce the Fermi coupling constant into the matrix element. Carefully write down the four spinors, using the facts that the electron and muon are at rest, and the muon is an antiparticle spinor. The neutrino–antineutrino are not at rest, but they are massless Dirac particle–antiparticle. Show that only γ^0 contributes to the matrix element, and use $\bar{u}\gamma^0 = u^\dagger$ to simplify the equation. Write down the 4×4 matrix $(1 - \gamma^5)$, and show that at each vertex the lepton and its neutrino have parallel spins, and that the ν_e is left handed, and the $\bar{\nu}_\mu$ right handed, so the muonium is in a triplet spin state. You knew that this had to be, but it is fun to see it actually work. Substitute the matrix element into the proper form of Equation (9.28), calculate $\langle B|B \rangle$. The muonium wave function is the same as hydrogen for our purposes. Evaluate the ratio of transition rates:

$$\frac{\Gamma(\mu^+ e^- \to \bar{\nu}_\mu \nu_e)}{\Gamma(\mu^+ \to e^+ \bar{\nu}_\mu \nu_e)} = \left(\frac{\alpha m_e}{m_\mu}\right)^3 \times 12\pi. \tag{9.161}$$

This is a very small number, but an instructive exercise.

Solution 9.7. Muonium annihilation $\mu^+ + e^- \to \nu_\mu + \nu_e$. This is similar to the decay of singlet positronium, only through the weak force rather than the electromagnetic. The branching fraction is very small, because m_μ^5 in the normal muon decay $\mu^+ \to e^+ + \bar{\nu}_\mu + \nu_e$ is replaced by $m_\mu^2 m_e^3 \alpha^3$, with $m_e^3 \alpha^3$ coming from $|\psi(0)|^2$, the hydrogen ground state wave function. There is only one diagram - t channel W exchange. The e^- emits a W^- and becomes a ν_e. The μ^+ absorbs the

W^- to become a $\bar{\nu}_\mu$. Or the other way around. Only one diagram is simpler than positronium annihilation, but the weak force has $(1 - \gamma^5)$, which makes up for it! The T matrix amplitude is:

$$-i\langle \nu_e \bar{\nu}_\mu | T | e^- \mu^+ \rangle = \bar{u}(p_3) \frac{-ig}{\sqrt{2}} \gamma^\mu \frac{1 - \gamma^5}{2} u(p_1) \frac{-i(g_{\mu\nu} + q_\mu q_\nu / M_W^2)}{q^2 - M_W^2}$$

$$\times \bar{v}(p_2) \frac{-ig}{\sqrt{2}} \gamma^\nu \frac{1 - \gamma^5}{2} v(p_4).$$

$$p_1 = (m_e, 0, 0, 0); \quad p_2 = (m_\mu, 0, 0, 0);$$

$$p_3 = (m_\mu/2, 0, 0, m_\mu/2); \quad p_4 = (m_\mu/2, 0, 0, -m_\mu/2);$$

$$q^2 = (p_1 - p_3)^2 = \frac{(m_e - m_\mu)^2}{4} - \frac{(m_e + m_\mu)^2}{4}$$

$$= -m_e m_\mu; \quad m_e m_\mu \ll M_W^2.$$

In the total energy, we may ignore the electron mass, and set $q^2 = 0$. The electron spinors are:

$$u^{(1)}(p_1) = \sqrt{2m_e} \begin{pmatrix} 1 \\ 0 \\ 0 \\ 0 \end{pmatrix}; \text{ and } u^{(2)}(p_1) = \sqrt{2m_e} \begin{pmatrix} 0 \\ 1 \\ 0 \\ 0 \end{pmatrix}.$$

The muon spinors are:

$$\bar{v}^{(1)}(p_2) = \sqrt{2m_\mu} \begin{pmatrix} 0 & 0 & 0 & -1 \end{pmatrix}; \text{ and }$$

$$\bar{v}^{(2)}(p_2) = \sqrt{2m_\mu} \begin{pmatrix} 0 & 0 & -1 & 0 \end{pmatrix}.$$

Next comes the electron neutrino:

$$\bar{u}^{(1)}(p_3) = \sqrt{m_\mu/2} \begin{pmatrix} 1 & 0 & -1^{\scriptscriptstyle \backprime} & 0 \end{pmatrix}; \text{ and }$$

$$\bar{u}^{(2)}(p_3) = \sqrt{m_\mu/2} \begin{pmatrix} 0 & 1 & 0 & 1 \end{pmatrix}.$$

Last but not least the muon anti-neutrino:

$$v^{(1)}(p_4) = \sqrt{m_\mu/2} \begin{pmatrix} 0 \\ 1 \\ 0 \\ 1 \end{pmatrix}; \text{ and } v^{(2)}(p_4) = \sqrt{m_\mu/2} \begin{pmatrix} -1 \\ 0 \\ 1 \\ 0 \end{pmatrix}.$$

The transition amplitude is:

$$\langle \nu_e \bar{\nu}_\mu | T | e^- \mu^+ \rangle = \frac{G_F}{\sqrt{2}} \bar{u}(p_3)\gamma^\mu(1 - \gamma^5)u(p_1) \times \bar{v}(p_2)\gamma_\mu(1 - \gamma^5)v(p_4).$$

In our representation

$$1 - \gamma^5 = \begin{pmatrix} 1 & 0 & -1 & 0 \\ 0 & 1 & 0 & -1 \\ -1 & 0 & 1 & 0 \\ 0 & -1 & 0 & 1 \end{pmatrix} ; \gamma^0 = \begin{pmatrix} 1 & 0 & 0 & 0 \\ 0 & 1 & 0 & 0 \\ 0 & 0 & -1 & 0 \\ 0 & 0 & 0 & -1 \end{pmatrix}$$

$$\gamma^3 = \begin{pmatrix} 0 & 0 & 1 & 0 \\ 0 & 0 & 0 & -1 \\ -1 & 0 & 0 & 0 \\ 0 & 1 & 0 & 0 \end{pmatrix} .$$

The list includes γ^3 because it makes a contribution through the product $\gamma^3\gamma^5$ that gives diagonal terms. The statement in the text that only γ^0 contributes is a mistake. However, if you worked it that way, you are only off by a factor of two in the amplitude. γ^1 and γ^2 cancel. For the electron spinor:

$$(1 - \gamma^5)u^{(1)}(p_1) = \sqrt{2m_e} \begin{pmatrix} 1 \\ 0 \\ -1 \\ 0 \end{pmatrix} ; (1 - \gamma^5)u^{(2)}(p_1) = \sqrt{2m_e} \begin{pmatrix} 0 \\ 1 \\ 0 \\ -1 \end{pmatrix} .$$

For the muon anti-neutrino spinor:

$$(1 - \gamma^5)v^{(1)}(p_4) = 0; \ (1 - \gamma^5)v^{(2)}(p_4) = 2\sqrt{m_\mu/2} \begin{pmatrix} -1 \\ 0 \\ 1 \\ 0 \end{pmatrix}$$

The first product vanishes because the right-handed anti-neutrino is going in the $-\hat{z}$ direction. Matrix multiplication gives the following elements:

$$\mu = 0 : \bar{u}^{(1)}(p_3)\gamma^0(1 - \gamma^5)u^{(1)}(p_1) = 0; \ \text{left handed neutrino};$$

$$\bar{u}^{(2)}(p_3)\gamma^0(1 - \gamma^5)u^{(2)}(p_1) = 2\sqrt{m_e m_\mu}.$$

There are no elements with different spin indices for electron and neutrino. The muon term is:

$$\bar{v}^{(2)}(p_2)\gamma^0(1-\gamma^5)v^{(2)}(p_4) = 2m_\mu; \quad \bar{u}(p_3)\gamma^0(1-\gamma^5)u(p_1)$$
$$\times \bar{v}(p_2)\gamma^0(1-\gamma^5)v(p_4) = 4m_\mu\sqrt{m_e m_\mu}.$$

After changing the sign of the γ^3 term because of the metric tensor $(\gamma^\mu\gamma_\mu = (\gamma^0)^2 - \vec{\gamma}^2)$ the result is the same, doubling the answer:

$$\sum_{\text{spins}} \bar{u}(p_3)\gamma^\mu(1-\gamma^5)u(p_1) \times \bar{v}(p_2)\gamma_\mu(1-\gamma^5)v(p_4) = 8m_\mu\sqrt{m_e m_\mu}.$$

This answer is a constant, independent of q^2, as it was for positronium annihilation. Substitute it for the positronium amplitude in Equation (9.23):

$$|B\rangle = 8\frac{G_F}{\sqrt{2}}m_\mu\sqrt{m_e m_\mu}\sqrt{2m_\mu}\int \frac{d^3p}{(2\pi)^3}f(p)\frac{1}{\sqrt{2m_e}}\frac{1}{\sqrt{2m_\mu}}$$

$$= 4G_F\psi(0)(m_\mu)^{3/2}.$$

$$\langle B|B\rangle = 16G_F^2 m_\mu^3|\psi(o)|^2.$$

Divide by four to average over the initial spins, and substitute it into the standard decay rate formula:

$$d\Gamma = 4G_F^2 m_\mu^3|\psi(0)|^2\frac{1}{2m_\mu}\frac{(2\pi)^4}{(2\pi)^6}\delta(\omega + \omega' - m_\mu)\delta(\vec{k}+\vec{k}')\frac{d^3k}{2\omega}\frac{d^3k'}{2\omega'}.$$

The δ functions get the usual treatment, and $|\psi(0)|^2 = (m_e\alpha)^3/\pi$ in our units gives:

$$\Gamma(\mu^+e^- \to \nu_e\bar{\nu}_\mu) = \frac{G_F^2 m_\mu^2 m_e^3\alpha^3}{4\pi^2}.$$

Finally, the ratio to muon decay:

$$\frac{\Gamma(\mu e \to \nu\bar{\nu})}{\Gamma(\mu \to e\nu\bar{\nu})} = 48\pi\alpha^3\left(\frac{m_e}{m_\mu}\right)^3.$$

This is a factor of four larger than the answer in the text because of the inclusion of the γ^3 contribution.

Problem 9.8. Do a literature search for experimental evidence for mesons and baryons containing more than one heavy quark. Start with the Particle Data Group and references therein. This is an active field of research, with new results from the LHC detectors every year.

Solution 9.8. This should be current results, so any solution is time stamped. Now (March, 2022) some interesting new double heavy quark particles are the baryon Ξ_{cc}^{++} at 3621.55 MeV mass, and life-time of 0.25×10^{-12} sec, quark content (u, c, c); and the two heavy mesons $B_c^+ \to J/\psi \pi^+$ at 6274 MeV, lifetime 0.5×10^{-12} sec, and the $B_c(2S)$ with the same quark content $(c\bar{b})$ that is 600 MeV heavier (6871.2 MeV), and decays by strong interaction into $B_c(1S) + 2\pi$. The $B_c(2S)$ is thought to be a psuedoscalar, although most excited states of $q\bar{q}$ pairs are vectors, like the rho, the K^*, and the quarkonium states.

Problem 9.9. Verify Equation (9.39), and try to fit it to $1^3S - 2^3S - 3^3S$ splittings for Υ and ψ families. Work in MeV-fermi units. $\hbar c = 197$ MeV-fm, and let $r_0 = 0.5$ fm, for the radius where the potential crosses the axis, although you do not really need either number to calculate energy level differences. The two families would be identical if Equation (9.39) were correct, and if $\Upsilon(3S) - \Upsilon(1S) = 895$ MeV, then the prediction for $\Upsilon(3S) - \Upsilon(2S)$ is about 20 MeV low, not too bad. You have only one free parameter, the constant C in the potential $V(r) = C \ln(r/r_0)$.

Solution 9.9. Logarithmic potential in WKB approximation. This analysis is outlined in the text. The three-dimensional formula for the WKB approximation with $l = 0$ is:

$$\int_0^{r_c} dr (E - C \ln(r/r_0))^{1/2} = \frac{\pi \hbar (n - 1/4)}{\sqrt{2m}}.$$

The constant C has dimensions of energy and is the strength of the potential. $V(r_0) = 0$, and $V(r_c) = E$. The definite integral can be evaluated by a change of variables. Let

$$y = E - C \ln(r/r_0); \quad dy = -C \frac{dr}{r}; \quad r = r_0 e^{(E-y)/C};$$

$$r = r_c \to y = 0; \quad \text{and } r = 0 \to y = \infty.$$

Reverse the limits and cancel the minus sign to get the integral:

$$\frac{1}{C}e^{E/C}\int_0^\infty e^{-y/C}\sqrt{y}\,dy = \frac{\pi\hbar(n-1/4)}{r_0\sqrt{2m}}; \text{ and } \int_0^\infty dt\sqrt{t}e^{-t} = \Gamma(3/2)$$

$$= 1/2\Gamma(1/2) = \frac{\sqrt{\pi}}{2},$$

where by definition $\Gamma(z) = \int_0^\infty t^{z-1}e^{-t}dt$. $\Gamma(z+1) = z\Gamma(z)$. $\Gamma(1/2)$ is easy to evaluate:

$$\Gamma(1/2) = \int_0^\infty e^{-t}\frac{dt}{\sqrt{t}}; \text{ let } t = x^2; \; dt = 2xdx;$$

$$\Gamma(1/2) = 2\int_0^\infty e^{-x^2}dx = \sqrt{\pi}.$$

The change of variables gives a standard Gaussian integral. Assembling everything gives:

$$e^{E/C} = \sqrt{\frac{2\pi C}{mc^2}} \times \frac{\hbar c(n-1/4)}{r_0 C};$$

Each of the two factors is dimensionless. The energy is the sum of two logarithms in agreement with Equation (9.39):

$$E = C(\ln(\hbar c(n-1/4)/(r_0 C)) + \frac{1}{2}\ln(2\pi C/(mc^2))).$$

In calculating energy level differences, only $\ln(n-1/4)$ matters. Everything else cancels. The numbers in the table are all calculated using the Υ $n=3$ to $n=1$ energy difference to calculate $C = 688$ MeV. This number for the charmonium family is $C = 724$ MeV, about a 5% difference, and that changes the numbers slightly in the ψ $log(r)$ column. For the Υ the only prediction is where the $n=2$ level lies, but for the ψ all of the differences are predictions. The agreement is not that bad — better than 10% in most cases. Remember that the mass scale differs by a factor of three. There are small rounding errors; 581+310=891, not 895.

Table logarithmic potential model of Υ and ψ ^3S states

	ΔE	Υ exp data	Υ $log(r)$	ψ exp data	ψ $log(r)$
$n = 3$ to $n = 1$	895 MeV	895 MeV	942 MeV	895 MeV
$n = 3$ to $n = 2$	332 MeV	310 MeV	353 MeV	310 MeV
$n = 2$ to $n = 1$	563 MeV	581 MeV	589 MeV	581 MeV

Problem 9.10. Verify the various relations listed in Equation (9.68). In particular obtain the formulas for p^2 and q^2.

Solution 9.10. The Lee, Oehme, and Yang expressions for the eigenvalues and eigenfunctions of the mass matrix.

$$(iM + \Gamma/2)\psi_\pm = \lambda_\pm\psi_\pm; \quad \psi_\pm = \frac{1}{\sqrt{|p|^2 + |q|^2}} \begin{pmatrix} p \\ \pm q \end{pmatrix}$$

$$\lambda_\pm = iM_{11} + \Gamma_{11}/2 \pm (pq);$$

then

$$p^2 = \frac{\Gamma_{12}}{2} + iM_{12}; \text{ and } q^2 = \frac{\Gamma_{12}^*}{2} + iM_{12}^*.$$

Substitute the state vector into the eigenvalue equation:

$$(iM_{11} + \Gamma_{11}/2)p \pm (iM_{12} + \Gamma_{12}/2)q = \lambda_\pm p; \text{ and}$$

$$(iM_{12}^* + \Gamma_{12}^*/2)p \pm (iM_{11} + \Gamma_{11}/2)q = \pm\lambda_\pm q.$$

Choose the plus sign, and substitute for λ_+:

$$(iM_{11} + \Gamma_{11}/2)p + (iM_{12} + \Gamma_{12}/2)q = (iM_{11} + \Gamma_{11}/2)p + p^2q;$$

it follows that $p^2 = iM_{12} + \Gamma_{12}/2$. Take the second equation, and choose the plus sign again:

$$(iM_{12}^* + \Gamma_{12}^*/2)p + (iM_{11} + \Gamma_{11}/2)q = (iM_{11} + \Gamma_{11}/2)q + pq^2;$$

now it follows that $q^2 = iM_{12}^* + \Gamma_{12}^*/2$.

Problem 9.11. Verify Equation (9.45) for $\vec{\sigma} \cdot \vec{r}/r$ in terms of spherical harmonics, and use this form to show that $\langle s = 0| S_{12} |s = 0\rangle = 0$, $\langle s = 1, m_s = \pm 1| S_{12} |s = 1, m_s = \pm 1\rangle = 3\cos^2\theta - 1$, and $\langle s = 1, m_s = 0| S_{12} |s = 1, m_s = 0\rangle = 2(1 - 3\cos^2\theta)$.

Solution 9.11. Derivation of Equation (9.45).

$$\frac{\vec{\sigma} \cdot \vec{r}}{r} = \sin\theta\cos\phi\sigma_x + \sin\theta\sin\phi\sigma_y + \cos\theta\sigma_z;$$

$$\sigma_x = \sigma_+ + \sigma_-; \quad \sigma_y = i(\sigma_- - \sigma_+).$$

Combining terms:

$$\frac{\vec{\sigma} \cdot \vec{r}}{r} = \sin\theta e^{-i\phi}\sigma_+ + \sin\theta e^{i\phi}\sigma_- + \cos\theta\sigma_z.$$

If you are just interested in $l = 0$ states, you can leave it in this form, but the spherical harmonics are more convenient for $l = 1$ P states, so substitute the definitions of the $l = 1$ spherical harmonics:

$$Y_1^{+1}(\theta, \phi) = -\sqrt{\frac{3}{8\pi}} \sin \theta e^{i\phi}; \quad Y_1^{-1}(\theta, \phi) = \sqrt{\frac{3}{8\pi}} \sin \theta e^{-i\phi};$$

$$Y_1^0(\theta) = \sqrt{\frac{3}{4\pi}} \cos \theta;$$

to obtain Equation (9.45):

$$\frac{\vec{\sigma} \cdot \vec{r}}{r} = \sqrt{\frac{4\pi}{3}} (\sqrt{2}(Y_1^{-1}\sigma_+ - Y_1^{+1}\sigma_-) + Y_1^0\sigma_z)$$

$$= \sqrt{\frac{4\pi}{3}} \begin{pmatrix} Y_1^0 & \sqrt{2}Y_1^{-1} \\ -\sqrt{2}Y_1^{+1} & -Y_1^0 \end{pmatrix}.$$

The tensor force term as written in Equation (9.44) is:

$$S_{12} = \frac{3(\vec{\sigma}_1 \cdot \vec{r})(\vec{\sigma}_2 \cdot \vec{r})}{r^2} - \vec{\sigma}_1 \cdot \vec{\sigma}_2.$$

The second part is easy:

$$-\vec{\sigma}_1 \cdot \vec{\sigma}_2 = -2S(S+1) + 3 = 3 \ (S = 0); \ = -1 \ (S = 1).$$

So for singlet spin states:

$$\langle S = 0 | \, S_{12} \, | S = 0 \rangle = 3 \langle S = 0 | \, \vec{\sigma}_1 \cdot \hat{r} \vec{\sigma}_2 \cdot \hat{r} \, | S = 0 \rangle + 3.$$

The operator for the first term is:

$$3\vec{\sigma}_1 \cdot \hat{r}\vec{\sigma}_2 \cdot \hat{r} = 4\pi (\sqrt{2}(Y_1^{-1}\sigma_{1+} - Y_1^{+1}\sigma_{1-}) + Y_1^0\sigma_{1z})$$

$$\times (\sqrt{2}(Y_1^{-1}\sigma_{2+} - Y_1^{+1}\sigma_{2-}) + Y_1^0\sigma_{2z}).$$

Multiplying it all out gives:

$$3\vec{\sigma}_1 \cdot \hat{r}\vec{\sigma}_2 \cdot \hat{r} = 4\pi \left\{ 2((Y_1^{-1})^2\sigma_{1+}\sigma_{2+} + (Y_1^{+1})^2\sigma_{1-}\sigma_{2-} \right.$$

$$- Y_1^{+1}Y_1^{-1}(\sigma_{1+}\sigma_{2-} + \sigma_{1-}\sigma_{2+}))$$

$$+ \sqrt{2}(Y_1^{-1}Y_1^0(\sigma_{1+}\sigma_{2z} + \sigma_{1z}\sigma_{2+})$$

$$- Y_1^{+1}Y_1^0(\sigma_{1-}\sigma_{2z} + \sigma_{1z}\sigma_{2-})) + (Y_1^0)^2\sigma_{1z}\sigma_{2z} \right\}.$$

The spin operators for different particles commute, so the order of writing is not important. Operate on the crossed spin state:

$$3\vec{\sigma}_1 \cdot \hat{r}\vec{\sigma}_2 \cdot \hat{r} \left|\uparrow\downarrow\right\rangle = 4\pi \left\{ -2Y_1^{+1}Y_1^{-1} \left|\downarrow\uparrow\right\rangle + \sqrt{2}(Y_1^{-1}Y_1^0 \left|\uparrow\uparrow\right\rangle \right.$$
$$\left. +Y_1^{+1}Y_1^0 \left|\downarrow\downarrow\right\rangle) - (Y_1^0)^2 \left|\uparrow\downarrow\right\rangle \right\}.$$

The opposite spin ket is very similar.

$$3\vec{\sigma}_1 \cdot \hat{r}\vec{\sigma}_2 \cdot \hat{r} \left|\downarrow\uparrow\right\rangle = 4\pi \left\{ -2Y_1^{+1}Y_1^{-1} \left|\uparrow\downarrow\right\rangle + \sqrt{2}(Y_1^{-1}Y_1^0 \left|\uparrow\uparrow\right\rangle \right.$$
$$\left. +Y_1^{+1}Y_1^0 \left|\downarrow\downarrow\right\rangle) - (Y_1^0)^2 \left|\downarrow\uparrow\right\rangle \right\}.$$

Combining the two into a singlet state and dropping the parallel spin states that are orthogonal gives:

$$3\vec{\sigma}_1 \cdot \hat{r}\vec{\sigma}_2 \cdot \hat{r} \left|S=0, m_s=0\right\rangle = 4\pi(2Y_1^{+1}Y_1^{-1} - (Y_1^0)^2) \left|S=0, m_s=0\right\rangle.$$

The spherical harmonic expression equals -3, canceling the $-\vec{\sigma}_1 \cdot \vec{\sigma}_2$ term, so that

$$\langle S=0, m_s=0 | S_{12} | S=0, m_s=0 \rangle = 0.$$

For the triplet states $\left|\uparrow\uparrow\right\rangle$ and $\left|\downarrow\downarrow\right\rangle$ only the $(Y_1^0)^2$ term in the operator contributes, giving:

$$\langle S=1, m_s=\pm1 | S_{12} | S=1, m_s=\pm1 \rangle = 3\cos^2\theta - 1.$$

This is the unnormalized $l=2$ spherical harmonic Y_2^0, so the tensor force can couple $l=0$ and $l=2$ states. Last but not least is the sum term for $\left|S=1, m_s=0\right\rangle$:

$$3\vec{\sigma}_1 \cdot \hat{r}\vec{\sigma}_2 \cdot \hat{r} \left|S=1, m_s=0\right\rangle$$
$$= 4\pi(-2Y_1^{+1}Y_1^{-1} - (Y_1^0)^2) \left|S=1, m_s=0\right\rangle$$
$$= 3(1 - 2\cos^2\theta) \left|S=1.m_s=0\right\rangle.$$

Giving

$$\langle S=1, m_s=0 | S_{12} | S=1, m_s=0 \rangle = 2(1 - 3\cos^2\theta).$$

Problem 9.12. Plot Equation (9.82) for the interference between regenerated $K_S \to \pi^+\pi^-$ and CP violating $K_L \to \pi^+\pi^-$. Estimate the relative normalization of the interference term from the experimental plot Figure 9.11. On the same graph plot the curve for an anti-carbon regenerator, demonstrating clearly how matter versus antimatter can be determined.

Solution 9.12. Interference between regenerated and CP violating $K^0 \to \pi^+\pi^-$. The equation should be (9.83), not (9.82). Apologies for the mistake. The correct equation is:

$$I(K \to \pi^+\pi^-) = C(|\rho|^2 e^{-\Gamma_S \tau} + |\eta_\pm|^2 e^{-\Gamma_L \tau}$$
$$+ 2|\rho||\eta_\pm| e^{-(\Gamma_S + \Gamma_L)\tau/2} \cos(\Delta m \tau + \phi_\rho - \phi_\pm)).$$

Some of the numbers are known. Decay rates $\Gamma_{K_s} = \hbar/\tau_s = 7.3 \times 10^{-6}$ eV; $\Gamma_L = \hbar/\tau_L = 1.3 \times 10^{-8}$ eV; $|\eta_\pm| = 2.3 \times 10^{-3}$; and $\Delta m = 0.47\ \Gamma_s = 3.4 \times 10^{-6}$ eV; $\Delta m/\hbar = 1.1 \times 10^{10}$ sec^{-1}. The phases have been measured, and we may use $\phi_\rho - \phi_\pm = -\pi/2$. In Figure 2 the overall normalization constant was derived from Curve C in Figure 9.11 at t=0, and the regeneration amplitude used was $|\rho|^2 = 1.4 \times 10^{-3}$, taken from Curve C at $t = 10 \times 10^{-10}$ sec. The plot speaks for itself. To get the red curve, I reversed the sign of the interference term in the formula.

Problem 9.13. The Wolfenstein form for the CKM matrix to order λ^5 for imaginary terms is[a]:

V_{CKM}

$$= \begin{pmatrix} 1 - \lambda^2/2 & \lambda & A\lambda^3(\rho - i\eta + i\eta\lambda^2/2) \\ -\lambda & 1 - \lambda^2/2 - i\eta A^2\lambda^4 & A\lambda^2(1 + i\eta\lambda^2) \\ A\lambda^3(1 - \rho - i\eta) & -A\lambda^2 & 1 \end{pmatrix}.$$

(9.162)

Note that now V_{cs} and V_{cb} have acquired complex phases that are order λ^4, and a very small part $\sim \lambda^5$ has been added to V_{ub}. The three

---

[a]L. Wolfenstein, *Phys. Rev. Lett.* **51**, 1945 (1983).

K_S->ππ and K_L->ππ Interference

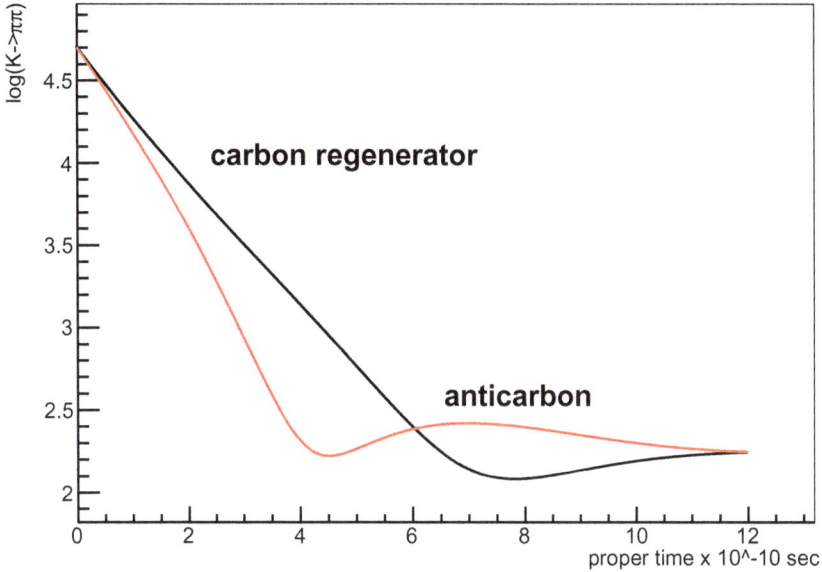

Figure 2: The formula in Figure 9.11 is reproduced in black. The y-axis is the common logarithm of $I(K \to \pi\pi)$. The red curve has the sign of the regeneration amplitude reversed for anti-carbon.

independent unitary triangles are Equations (9.99)–(9.101).

$$V_{ub}^* V_{ud} + V_{cb}^* V_{cd} + V_{tb}^* V_{td} = 0,$$

$$V_{ud}^* V_{us} + V_{cd}^* V_{cs} + V_{td}^* V_{ts} = 0,$$

$$V_{us}^* V_{ub} + V_{cs}^* V_{cb} + V_{ts}^* V_{tb} = 0.$$

Top is the golden triangle. The other two are flat in Wolfenstein's $\sim\lambda^3$ approximation. Use the more accurate form Equation (9.162) to plot all three triangles in a two dimensional complex space. Show that the second and third triangles, like the first, are almost exactly right angle triangles, and that all three have the same area.

Solution 9.13. Wolfenstein approximation to λ^5:

V_{CKM}

$$= \begin{pmatrix} 1 - \lambda^2/2 & \lambda & A\lambda^3(\rho - i\eta + i\eta\lambda^2/2) \\ -\lambda & 1 - \lambda^2/2 - i\eta A^2\lambda^4 & A\lambda^2(1 + i\eta\lambda^2) \\ A\lambda^3(1 - \rho - i\eta) & -A\lambda^2 & 1 \end{pmatrix}$$

This approximation is correct to order λ^5 in the imaginary part, but only to order λ^3 in the real part. Note that now V_{cs} and V_{cb} have acquired complex phases that are order λ^4, and a very small part $\sim \lambda^5$ has been added to V_{ub}. The three independent unitary triangles are Equations (9.99), (9.100), and (9.101).

$$V_{ub}^* V_{ud} + V_{cb}^* V_{cd} + V_{tb}^* V_{td} = 0;$$

$$V_{ud}^* V_{us} + V_{cd}^* V_{cs} + V_{td}^* V_{ts} = 0;$$

$$V_{us}^* V_{ub} + V_{cs}^* V_{cb} + V_{ts}^* V_{tb} = 0.$$

The top one is the golden triangle, with area $(A\lambda^3)^2 \eta/2$. The second one, using the improved Wolfenstein form, is:

$$\lambda(1 - \lambda^2/2 - (1 - \lambda^2/2 - i\eta A^2\lambda^4) - A^2\lambda^4(1 - \rho + i\eta)) = 0.$$

Notice that the term $A^2\lambda^4\rho$ is all by itself, and not canceled. So the real part terms smaller than λ^3 have to be set to zero. This gives the equation:

$$\lambda(1 - \lambda^2/2 - (1 - \lambda^2/2 - i\eta A^2\lambda^4) - i\eta A^2\lambda^4) = 0.$$

With λ factored out, this gives a triangle with sides $1 - \lambda^2/2, i\eta A^2\lambda^4$, and $1 - \lambda^2/2$, the extra term on the hypotenuse $\eta A^2\lambda^4$ being negligible. So the area of the triangle is $\lambda^2(1 - \lambda^2/2)\eta A^2\lambda^4/2 \sim \eta A^2\lambda^6/2$. The third unitarity relation proceeds in the same way. The triangles are isosceles, and approximately right triangles, because the opening angles are small.

Problem 9.14. Work out the kinematics for an asymmetric $e^- e^+$ collider tuned to the $\Upsilon(4S)$. Set the electron ring at $10\,\mathrm{GeV}$. Calculate the energy of the positron ring, and the velocity of the $\Upsilon(4S)$ in the laboratory. What is the maximum opening angle and spatial distance between the two vertices B and $\bar{B}$ if they decay simultaneously after traveling one lifetime?

Solution 9.14. Kinematics of an asymmetric $e^- e^+$ collider. Set the ring energies to match the $\Upsilon(4S)$ mass of 10579 MeV.

$$s = (p_1 + p_2)^2 = (E_1 + E_2)^2 - (\vec{p}_1 + \vec{p}_2)^2 = 4E_1 E_2; \quad E_1 = 10 \text{ GeV};$$

$$E_2 = 2.8 \text{ GeV}; \quad v_c = \frac{|\vec{p}_1| - |\vec{p}_2|}{E_1 + E_2} = 0.56.$$

We have neglected the electron and positron masses, assumed the colliding beams are in opposite directions, and the electron energy is 10 GeV. The center of mass moves in the electron direction with $v_c = 0.56$; $\gamma_c = 1.21$. The B^0 lifetime in its rest frame is $\tau_B = 1.5 \times 10^{-12}$ sec, giving a mean flight distance of $\gamma_c v_c c\tau_B = 300\mu$m. The velocity of either B in the center of mass, calculated in the text, is $v_0 = 0.061$, about 10% of the velocity in the laboratory, giving a maximum opening angle between the two B^0's in the laboratory of about 12^0. If each decays after one lifetime, each B travels 300 μm downstream, and the maximum separation between the vertices is 63μm in the transverse direction. In $B\bar{B}$ mixing, the second B travels on average 300 μm downstream before decaying.

Problem 9.15. Given that $\langle B|\bar{B}\rangle = 0$, and $\langle B|B\rangle = \langle\bar{B}|\bar{B}\rangle = 1$, show that Equations (9.123) and (9.124) are correctly normalized at $t = 0$.

Solution 9.15. At $t = 0$, Equation (9.123) becomes:

$$|B(0)\rangle = \frac{1}{2(1 + \epsilon_B)}(1 + \epsilon_B)2\,|B\rangle = |B\rangle\,.$$

The $|\bar{B}(0)\rangle$ argument is the same.

Problem 9.16. Verify the relation for the $B\bar{B}$ mass mixing matrix element $\text{Im}(M_{12}) = |M_{12}|\sin(2\beta)$.

Solution 9.16. Begin with the formula for ϵ in Equation (9.75):

$$\epsilon = \frac{p - q}{p + q} = \frac{\text{Im}(M_{12})}{2(|\Gamma_{12}|^2/4 + |M_{12}|^2)}(-|\Gamma_{12}|/2 + i|M_{12}|).$$

The $B\bar{B}$ mixing differs from $K\bar{K}$ because there are many B decay modes, but only a few that connect B and $\bar{B}$. The result is only one lifetime, and $\Gamma_{12} = 0$. This simplifies the formula for ϵ_B:

$$\epsilon_B = i\frac{\sin\phi_{12}}{2}; \text{ where } M_{12} = |M_{12}|e^{i\phi_{12}}.$$

ϵ_B is pure imaginary. The phase of M_{12} comes from the CKM matrix elements in the box diagram for $B\bar{B}$ mixing, Figure 9.13.

$$(V_{tb})^2(V_{td}^*)^2 = (A\lambda^3)^2(1 - \rho + i\eta)(1 - \rho + i\eta)$$
$$= A^2\lambda^6((1 - \rho)^2 - \eta^2 + 2i\eta(1 - \rho)); \text{ in Wolfenstein form.}$$

This gives

$$\tan(\phi_{12}) = \frac{2\eta(1 - \rho)}{(1 - \rho)^2 - \eta^2}; \text{ and } \sin(\phi_{12}) = \frac{2\eta(1 - \rho)}{(1 - \rho)^2 + \eta^2}.$$

Then Equation (9.111) shows that $\sin(\phi_{12}) = \sin(2\beta)$, where β is the right-hand opening angle in the "golden" triangle, Figure 9.12.

Problem 9.17. Show that the numerator and denominator for Equation (9.127) are

$$|\langle f|B(t)\rangle|^2 - |\langle f|\bar{B}(t)\rangle|^2 = \frac{|\langle f|B\rangle|^2 e^{-\Gamma t}}{|1 + \epsilon_B|^2} 4\text{Re}(\epsilon_B e^{-i\Delta mt}),$$

(9.163)

and

$$|\langle f|B(t)\rangle|^2 + |\langle f|\bar{B}(t)\rangle|^2 = \frac{|\langle f|B\rangle|^2 e^{-\Gamma t}}{|1 + \epsilon_B|^2} 2(1 + |\epsilon_B|^2).$$

(9.164)

Solution 9.17. Begin with Equations (9.123) and (9.124):

$$|B(t)\rangle = \frac{e^{-\Gamma t/2}}{2(1 + \epsilon_B)}((1 + \epsilon_B)(1 + e^{-i\Delta mt})|B\rangle + (1 - \epsilon_B)(1 - e^{-i\Delta mt})|\bar{B}\rangle);$$

and

$$|\bar{B}(t)\rangle = \frac{e^{-\Gamma t/2}}{2(1 - \epsilon_B)}((1 + \epsilon_B)(1 - e^{-i\Delta mt})|B\rangle + (1 - \epsilon_B)(1 + e^{-i\Delta mt})|\bar{B}\rangle).$$

The final state $|f\rangle$ is accessible to both B^0 and $\bar{B}^0$. If $|f\rangle$ is CP even, we assume that the two amplitudes are the same $\langle f|B\rangle = \langle f|\bar{B}\rangle$. The CP violation will then give different time dependences.

$$\langle f|B(t)\rangle = \frac{e^{-\Gamma t/2}\langle f|B\rangle}{1 + \epsilon_B}(1 + \epsilon_B e^{-i\Delta mt});$$

and

$$\langle f|\bar{B}(t)\rangle = \frac{e^{-\Gamma t/2}\langle f|B\rangle}{1 + \epsilon_B}(1 - \epsilon_B e^{-i\Delta mt});$$

Then

$$|\langle f|B(t)\rangle|^2 = \frac{e^{-\Gamma t}\langle f|B\rangle|^2}{1 + |\epsilon_B|^2}(1 + \epsilon_B e^{-i\Delta mt})(1 + \epsilon_B^* e^{i\Delta mt});$$

where $\epsilon_B = i\sin(2\beta)/2$ is pure imaginary, giving:

$$|\langle f|B(t)\rangle|^2 = \frac{e^{-\Gamma t}\langle f|B\rangle|^2}{1 + |\epsilon_B|^2}(1 + \sin(2\beta)\sin(\Delta mt) + \sin^2(2\beta)/4);$$

and

$$|\langle f|\bar{B}(t)\rangle|^2 = \frac{e^{-\Gamma t}|\langle f|B\rangle|^2}{1+|\epsilon_B|^2}(1 - \sin(2\beta)\sin(\Delta mt) + \sin^2(2\beta)/4).$$

The time-dependent asymmetry is the difference divided by the sum:

$$\frac{|\langle f|B(t)\rangle|^2 - |\langle f|\bar{B}(t)\rangle|^2}{|\langle f|B(t)\rangle|^2 + |\langle f|\bar{B}(t)\rangle|^2} = \frac{\sin(2\beta)\sin(\Delta mt)}{1+\sin^2(2\beta)/4};$$

in agreement with Equation 9.127.

Problem 9.18. Show that if $\langle f|B\rangle = -\langle f|\bar{B}\rangle$ a minus sign appears in Equation (9.127).

Solution 9.18. This one is easy. If $\langle f|\bar{B}\rangle = -\langle f|B\rangle$ the sign on ϵ_B in Equations (9.125) and (9.126) flip, and that changes the sign of Equation (9.127).

www.ingramcontent.com/pod-product-compliance
Lightning Source LLC
Chambersburg PA
CBHW060306220326
41598CB00027B/4247